rspb a home

Woodpeckers

Gerard Gorman

BLOOMSBURY WILDLIFE

LONDON · OXFORD · NEW YORK · NEW DELHI · SYDNEY

BLOOMSBURY WILDLIFE
Bloomsbury Publishing Plc
50 Bedford Square, London, WC1B 3DP, UK

BLOOMSBURY, BLOOMSBURY WILDLIFE and the Diana logo are trademarks of
Bloomsbury Publishing Plc

First published in Great Britain 2018

A catalogue record for this book is available from the British Library

Library of Congress Cataloguing-in-Publication data has been applied for

ISBN: PB: 978-1-4729-5118-2; ePub: 978-1-4729-5119-9; ePDF: 978-1-4729-5117-5

2 4 6 8 10 9 7 5 3 1

Design by Rod Teasdale

Printed in China by C&C Offset Printing Co., Ltd.

To find out more about our authors and books visit www.bloomsbury.com
and sign up for our newsletters

For all items sold, Bloomsbury Publishing will donate a minimum of 2% of the publisher's receipts
from sales of licensed titles to RSPB Sales Ltd, the trading subsidiary of the RSPB. Subsequent sellers
of this book are not commercial participators for the purpose of Part II of the Charities Act 1992.

FEB 0 8 2019

Contents

Meet the Woodpeckers

Woodpeckers are amazing birds. They have one of the most extraordinary anatomies in the animal world, every aspect of which is pitched towards working on wood. This enables them to make holes in trees and communicate with each other by mechanical means, drumming on timber with their bills. Quite simply, woodpeckers have honed their carpentry skills to levels no other birds can match. Their behaviour has long fascinated people across the globe and has led to them being admired as strong and hard-working animals and, in folklore, as the first 'tree surgeons' and 'guardians of the forest'.

Three species of woodpecker are resident in the British Isles – the Green Woodpecker *(Picus viridis)*, the Great Spotted Woodpecker *(Dendrocopos major)* and the Lesser Spotted Woodpecker *(Dryobates minor)*. A fourth, the Wryneck *(Jynx torquilla)*, is today only an occasional breeder in Britain, but regularly visits as a migrant. These four wonderful birds are the focus of this book, although we will also look at some of their remarkable relatives around the world.

The only other woodpecker to have ever officially occurred in Britain was an immature male Yellow-bellied Sapsucker *(Sphyrapicus varius)* that stayed on the island of Tresco in the Isles of Scilly between 25 September and 6 October 1975. This North American species is one of the few truly migratory woodpeckers, which probably explains its appearance in Britain as for some reason it diverted from its normal route southwards and crossed the Atlantic.

Above: The familiar Green Woodpecker is Britain's largest woodpecker species.

Opposite: Male Great Spotted Woodpecker – by far the most common and widespread woodpecker in Britain.

Left: The Wryneck is now a very rare breeding bird in Britain.

Right: Male Lesser Spotted Woodpecker – Europe's smallest woodpecker is sadly in decline in Britain.

What is a woodpecker?

Woodpeckers are small to medium-sized woodland birds. They are non-passerines; that is, not songbirds, clinging to tree trunks rather than perching on branches. A typical woodpecker (like the Great Spotted, the one that most people in Britain will be familiar with) has a distinct and familiar shape, being stocky, with a long, stout, pointed bill, a longish stiff tail and short strong legs with hooked claws. You can identify a Green, Great Spotted or Lesser Spotted Woodpecker with just one glimpse. That is not the case with the Wryneck, which, along with some other woodpeckers around the world, is 'atypical' – we will get back to this in the next chapter.

Below: With its stocky body, stout bill and stiff tail pressed against a tree trunk, the Great Spotted Woodpecker is instantly recognisable as a 'typical woodpecker'.

Mandela's Woodpecker

The earliest known woodpecker fossil from the African continent dates from the Pliocene. It was unearthed by a French–German team in South Africa and in 2012 the 5–3-million-year-old woodpecker it came from was named *Australopicus nelsonmandelai* as a tribute to Nelson Mandela on his 94th birthday. Fascinatingly, 'Mandela's Woodpecker' is regarded as having been more closely related to the woodpeckers found today in Europe and the Americas rather than those in Africa.

Right: Nelson Mandela received many honours during his lifetime, including having a woodpecker named after him!

In more scientific terms, woodpeckers are members of the Picidae, a family of birds in the Piciformes order. Studies on genetics (DNA sequence analysis) and morphology (structure and physique) suggest their closest relatives are the honeyguides of Africa and Asia, and barbets, a big family of birds found in Africa, Asia and South America. Toucans are a little more distantly related. The Piciformes began to evolve around 60 million years ago in the Paleocene epoch, which is after the mass dinosaur extinction. Woodpeckers split away from their relatives about 50 million years ago and would have done well in the vast forests that covered much of the land at that time. They have, therefore, been around a lot longer than we have – humans began to evolve from our ape-like ancestor between five and seven million years ago. The woodpeckers we see today are probably much like those that lived around five million years ago. The fossil record tells us that woodpeckers first evolved in what is now Europe and Asia. The earliest woodpecker fossil discovered, in central France, dates from the late Oligocene/early Miocene epoch, more than 25 million years ago. They then spread to Africa and the Americas, but never quite made it to Australia, New Zealand and Madagascar.

Woodpeckers in order

Above: Statue of the 'Father of Taxonomy' Carl Linnaeus. A Swedish physician, botanist and zoologist, he was also known as Carolus Linnaeus and Carl von Linné.

In the 18th century Carl Linnaeus, a Swedish zoologist, devised a system of classifying all living things. This is commonly referred to as the binomial system because the name of each organism is in two parts – the first indicates the genus to which the species belongs, the second the exact species. For example, the Green Woodpecker is *Picus viridis* (genus *Picus*, species *viridis*). Subspecies, or races, are differentiated by using a third name; for example, the Green Woodpecker that occurs in Britain and much of Europe is *Picus viridis viridis*, but the one in Italy and the Balkans is *Picus viridis karelini*.

The science of naming and putting animals in order that Linnaeus invented is called taxonomy. Unfortunately, it is often a little disordered, as taxonomists often disagree on how to apply it and even on what constitutes a species and what does not. Woodpecker taxonomy is particularly messy. Why is it important to mention this here? Well, because the taxonomic jumble, and the fact that we are not even sure whether some species still exist or not, means it is hard to decide just how many woodpecker species there are on our planet. So, depending upon which authority you follow, there are probably between 225 to 250 different species worldwide today.

Classification of Britain's woodpeckers

The woodpecker family (Picidae) is divided into three subfamilies: Jynginae (wrynecks), Picumninae (piculets) and Picinae ('true' woodpeckers). The four species in the British Isles fall into two of the above subfamilies, and into four genera. This is how the four perch in the taxonomic animal kingdom.

Kingdom: Animalia (animals)
Phylum: Chordata (vertebrates)
Class: Aves (birds)
Order: Piciformes (woodpeckers and allies)
Family: Picidae (woodpeckers)
Subfamily: Jynginae (wrynecks), Picinae (true woodpeckers)

Genus: *Jynx, Dendrocopos, Dryobates, Picus*
Species: *Jynx torquilla* (Wryneck), *Dendrocopos major* (Great Spotted Woodpecker), *Dryobates minor* (Lesser Spotted Woodpecker), *Picus viridis* (Green Woodpecker).

Sexual dimorphism

Most woodpeckers are sexually dimorphic, which in everyday terms means that males and females are different in size or appearance. The most common way in which the plumages of the sexes differ usually involves males having coloured areas on the crown or face – often red, sometimes yellow – which females lack. These differences are often negligible and can be difficult to judge, but in a few species sexual dimorphism is very pronounced. For example, the sexes of a North American woodpecker called Williamson's Sapsucker are so different (males being colourful, females plain) that in the past they were considered separate species. Of course, there are exceptions – male and female Wrynecks look virtually identical.

Above: Female (left) and male (right) Williamson's Sapsuckers look so unalike, you might be forgiven for thinking they were different species. In fact, it was once thought they were!

Meet the Green Woodpecker

Above and right: World and British Isles range maps of Green Woodpecker. Green indicates where it is resident all year round. Note that these are only guides and locally breeding birds will not occur over all the areas shown.

The Green Woodpecker is unlikely to be mistaken for any other British bird. It's easily the biggest woodpecker in Britain at 31–35cm (12-13.8in) long from bill-tip to tail-tip – about the size of a Jackdaw – and weighs around 190g (6.7oz). It's well-named, too, because it is mostly green, although its flight feathers are blackish, dotted with white spots, and it has a yellow rump which usually shows when it flies. Both sexes have a red cap and a black 'Lone Ranger' face mask. If you get a close look at its bill, you'll see that the upper mandible is dark and the lower yellowish. Males have a black malar stripe (often mistakenly called a moustache – a moustache starts above the bill, a malar starts from the lower mandible) with a red centre, but on females it is solid black, so a good look at the face is needed to determine a bird's sex. Young Green Woodpeckers differ from their parents until they undergo their first feather moult in autumn. They lack the black mask and are heavily scaled and spotted black and white.

Most taxonomists recognise three subspecies: *viridis* in Britain and continental Europe into Russia, *karelini* in Italy, the Balkans, the Caucasus and on to Turkmenistan, and *innominatus*, which is found only in the Zagros Mountains of Iran. However, the differences between these races are slight.

Most of the world's Green Woodpeckers live in Europe. Besides Britain (there are none in Ireland) they range from southern Scandinavia to the Mediterranean and eastwards as far as Asia Minor. Fortunately, Green Woodpeckers are not rare in Britain, at least in southern England. They have declined in parts of Wales, but in other areas, such as lowland Scotland, they have increased. They do not migrate; in fact they seldom move far, so hardly ever turn up outside their usual range.

Green Woodpeckers inhabit open deciduous and mixed woodlands, orchards, parks, gardens, heathland and dunes where there are some large trees for nesting and grasslands nearby where they can feed on ants – we'll get back to their liking for ants later. They have adapted to live on fields and lawns, which have increased greatly since Britain's wildwood was cleared. The disappearance of Green Woodpeckers is often due to the heavy use of herbicides and pesticides. Frequent ploughing in some areas can also result in a decline in numbers.

Above: Three Green Woodpeckers. From left to right, male (red in the black malar stripe), female (solid black malar stripe), juvenile (no black mask, heavily streaked, scaled and spotted).

Below: Green Woodpeckers on heathland (left) and in broadleaved woodland (right) – two habitats in Britain that the species often frequents.

Meet the Great Spotted Woodpecker

Above and below: World and British range maps of Great Spotted Woodpecker. Green indicates where it is resident all year round. Note that these are only guides and locally birds will not occur over all the areas shown.

Unlike the Green Woodpecker, the Great Spotted Woodpecker is not named that well as it is not very spotted. A better descriptive name would perhaps be 'pied' woodpecker, and in the past it was often called that. They are 21–23cm (8-9in) in length – the size of a Starling – and weigh around 85g (3oz). Both sexes are glossy black above, with two white oval 'shoulder' patches (formed by the outer scapulars and inner coverts). Underneath, from the throat to the belly they are white, sometimes buffy, and scarlet on the undertail. A white face is crossed by a black line (called the post-auricular stripe) which reaches the nape. The tail is mostly black, except for the outer feathers which are mostly white. The wings are black with five or six rows of white spots that form bars. The sexes differ in that males have a red nape patch. Both male and female juveniles have an all-red crown and their undertail is pinkish.

Depending on which checklist you follow, Britain is home to either its very own *anglicus* race or *pinetorum* which is also found across mainland Europe as far as Russia.

Above: Female Great Spotted Woodpecker – lacking the red nape patch.

Left: Male Great Spotted Woodpecker – showing the red nape patch.

The Great Spotted Woodpecker is the most widespread woodpecker in the world. From west to east, it is found from Ireland through Eurasia to Japan, and from north to south from the Arctic tree-line in Scandinavia to North Africa and even the Canary Islands. Most of the population stays put, but in some years irruptions, not true migrations, occur in response to a lack of food – this usually involves populations in cold climates heading west or south.

Below: Juvenile Great Spotted Woodpecker – showing an all-red crown and pink rather than red under the tail.

In Britain, they are doing very well. According to the British Trust for Ornithology (BTO), they have increased rapidly in recent decades and colonised Ireland in 2008. Ultimately, the best habitat for them is mature deciduous forests, where they reach their highest densities, but in Britain today they are found just about anywhere with trees – in conifer plantations, mixed forests, orchards, copses, parks and gardens.

Meet the Lesser Spotted Woodpecker

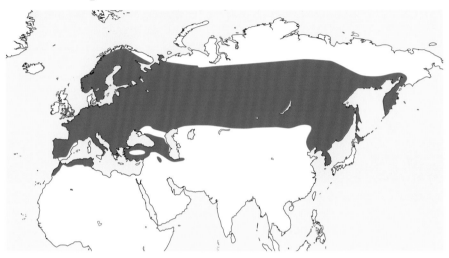

Above and below: World and British range maps of Lesser Spotted Woodpecker. Green indicates where it is resident all year round. Note that these are only guides and locally birds will not occur over all the areas shown.

The Lesser Spotted Woodpecker is mainly black and white, 'pied' like its larger relative the Great Spotted Woodpecker, but that is where the similarity ends. At just 14–16cm (5.5–6.3in) long it is much smaller – the size of a Chaffinch – and weighs around 21g (0.7oz). In fact, it is the smallest woodpecker in Europe. Both sexes are barred black and white, ladder-backed, from the mantle (upper back) to the rump. Underneath they are white with fine black streaks on the breast and flanks. Their flight feathers are barred black and white. The central tail is black, the outer tail feathers white with fine black bars. The face and throat are white and a black malar stripe runs to the sides of the neck where it meets another black stripe. The sexes differ in that males have red on their crown, which females lack. So female Lesser Spotted Woodpeckers are truly pied as they have no red feathers at all. Both juveniles are duller than the adults, often brownish and more finely streaked underneath, but males usually have a pink forecrown whilst females do not.

Those living in England and Wales are regarded as a distinct subspecies, *comminutus*. Elsewhere, across a vast

Left and above: Three Lesser Spotted Woodpeckers – male (with a red crown), female (which lacks red) and juvenile (browner, duller, less glossy plumage).

area from Britain southwards to North Africa and eastwards to Japan, there are another 10 to 12 subspecies – again depending on who you believe – which vary in plumage and size, but often only slightly. They seldom move far from where they breed, although extreme weather and food shortages sometimes cause minor irruptions.

Below: An adult male Lesser Spotted Woodpecker clinging to a thin upright branch in a classic pose.

Although they are doing fine globally, they are in big trouble in Britain. Numbers have fallen sharply since the 1980s and, worryingly, the reasons are still unclear, although the loss and degradation of the deciduous woodland they need is certainly involved. Recent studies have found that many nesting attempts fail and all British breeding records now need reporting to the Rare Breeding Birds Panel. They are essentially birds of boreal deciduous woodlands, but are found in mixed deciduous–coniferous forests, riverine woods and even man-made habitats like orchards and parks. In Britain, woods with mature oaks and traditionally managed orchards are important, but these have disappeared from many areas. A thing they really need are snags (dead or rotting dry branches) where they can feed, breed and drum.

Meet the Wryneck

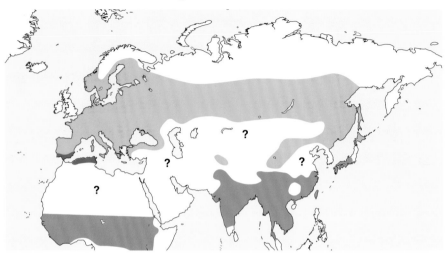

Above and below: World and British Isles range maps of Wryneck. Note that these are only guides and locally birds will not occur over all the areas shown. Green shows where the Wryneck is resident year-round; yellow shows where it is present during the summer breeding season; blue shows where it is present as a winter visitor; red shows areas that it visits at times of migration. On the British map, red is the area where Wrynecks usually occur when on passage.

The Wryneck (or Eurasian Wryneck, to give it its full name) doesn't look like a true woodpecker, and that's because it's not. It's a member of the woodpecker family, but lacks some of the 'typical' features that its cousins have, such as a stout bill and rigid tail. With its slim body, short rounded wings and fine bill, it is often said to look more like a songbird. They are 16–19cm (6.3–7.5in) long, with their elongated, rounded tail making up about a third of this, and weigh around 38g (1.3oz). They are cryptically patterned grey, brown and buff on their upperparts, with a dark brown band running from the crown onto the back and another across the ear-coverts. They are cream below with dark spots, streaks, bars and chevrons and the tail has four dark bands. When it comes to recognising the sexes, Wrynecks are very tricky, looking almost identical, although males are a touch more reddish-brown and yellowish underneath. Young birds look very like their parents, but are a little duller, with pale bars below rather than chevrons.

They love sunny woodlands, orchards, parks and gardens with open areas like glades, pastures, meadows and dunes, where there is short grass or bare ground with plenty of ants, their favourite food. In Britain, they often turn up on barren coastal headlands.

Most taxonomists recognise six subspecies, which look very similar. The *torquilla* race, which is found across most of Europe, occurs in Britain. They also differ from other European woodpeckers in being migratory. Most move south to Africa at the end of summer and return north to breed in spring, although many are also resident in places like southern Spain. Around 20 pairs bred in Scotland in the 1970s, but then they started to decline; nowadays only the odd pair are found nesting each year.

Above: When on the ground the camouflage colours of Wrynecks help them blend in with surrounding vegetation.

Below: Barren stony ground (left) and grazed pasture (right) – two typical foraging habitats that Wrynecks use when they pass through Britain.

Woodpeckers Around the World

Although this book focuses on British woodpeckers, it's important to place our woodpeckers in a global context. Woodpeckers are found on every continent except Australasia and Antarctica, and they are also missing from islands such as Greenland, Madagascar and Hawaii.

That woodpeckers are absent in the treeless terrains of the polar regions is understandable, but why haven't they ever pecked away in the seemingly suitable forests of Australia, New Zealand, New Guinea and Madagascar? Well, the explanation is that woodpeckers had not evolved in those areas of the ancient supercontinent of Gondwanaland that subsequently became those regions, before it broke up and the continents as we know them now drifted apart. Nevertheless, they are still widespread, having found habitat niches from sea level to high mountains, some even living above the tree-line. In terms of numbers of species, they are most abundant in tropical regions, with South America being a woodpecker paradise, closely followed by South East Asia. Next up is Africa although, considering its size, it has relatively few woodpeckers.

Below: A Middle Spotted Woodpecker with its red crown feathers raised. Though widespread on mainland Europe, this species has never been seen in Britain.

Opposite: Lesser Sri Lanka Flameback – just one of the many stunningly coloured woodpeckers found around the world.

The diversity of woodpeckers

Many woodpeckers are rather plain with green, black and pied plumages common for the family, though most also have patches of red or yellow. Yet there are some multicoloured species and others that have striking crests and they also vary greatly in size. The diminutive piculets, most of which are found in South America, are the smallest at just 8–9cm (3.1–3.5in) in length, the size of a Goldcrest. At the other end of the scale is the crow-sized Great Slaty Woodpecker of South East Asia which can reach 51cm (20.1in) from the tip of its bill to the end of its tail. The following are just a few examples from around the world to illustrate the rich diversity of the family in both appearance and behaviour.

What's in a name?

A perusal of books or an internet search for *Picus viridis* will soon turn up some other names for the bird known as Green Woodpecker in Britain, but don't panic. *Eurasian Green Woodpecker* or *European Green Woodpecker* are simply alternative names used to avoid confusion with the other 'green' woodpeckers around the world. Similarly, *Jynx torquilla* can be found as *Eurasian Wryneck* and *Northern Wryneck*, to differentiate it from the *Red-throated Wryneck* of Africa (which, by the way, also goes under the name of *Rufous-necked Wryneck* and *Rufous-breasted Wryneck!*). Furthermore, besides the wrynecks, some other woodpecker species are not even called 'woodpecker' in English. There are four sapsuckers in Central and North America, at least six flickers across the Americas, yellownapes and flamebacks in South East Asia, and many piculets, mostly in South America.

Flicker

Sapsucker

Yellownape

Flameback

Speckled-chested Piculet
Picumnus steindachneri

There are at least 27 species of
piculets in the genus *Picumnus*
(ongoing taxonomic studies may
reveal more) and all are tiny. The
Speckled-chested Piculet is restricted
to a few areas of mountain rainforest
with epiphytes, vines and bamboo
thickets in northern Peru. This little
gem is endangered as its habitats are
felled for timber, legally and illegally,
and cleared for plantations. Despite
their size, and small bills, many
piculets are avid drummers.

Great Slaty Woodpecker
Mulleripicus pulverulentus

The giant of the family at around half
a metre (about 20in) in length, the
Great Slaty lives in tropical lowland
forests and jungles with towering
trees, from the Himalayas through
Indochina to the Philippines and
Borneo. This whopping woodpecker
moves through the canopy in family
parties making distinctive whinnying
calls. Once quite common, this is
another rainforest species that is
under threat from deforestation.
Interestingly, despite its large bill, it is
not known to drum.

WOODPECKERS AROUND THE WORLD

Black Woodpecker
Dryocopus martius

This iconic bird is by far the biggest woodpecker in Europe, with some individuals reaching 50cm (19.7in) in length. Despite many rumours and reports, it has never officially managed to cross the sea from mainland Europe to Britain, although it is found close to the coast in France, the Low Countries and Scandinavia. It ranges over a vast area eastwards to China and Japan, and is often locally common. Black Woodpeckers are often vocal and both sexes drum loudly and powerfully.

Cuban Green Woodpecker
Xiphidiopicus percussus

There are many woodpeckers in the world that are basically green in colour and several have the word in their name. One such is the Cuban Green Woodpecker, which is endemic to that island. It is an adaptable woodpecker, common and widespread, inhabiting woodlands of all kinds including urban gardens. The Cuban Green calls frequently with one call replicated in a local name for the bird, *Jorre Jorre*. This species does not drum, however.

Yellow-fronted Woodpecker
Melanerpes flavifrons

This beautiful, colourful woodpecker is found in eastern South America, in parts of Brazil, Argentina and Paraguay. It is something of a food opportunist, an omnivore, eating insects, small reptiles, seeds and fruit. Conveniently, for photographers, it frequently visits wildlife-lodge bird tables, particularly where oranges and bananas have been placed out. A gregarious woodpecker, it roves through open woodlands, groves and orchards in boisterous, chattering family parties.

Blond-crested Woodpecker
Celeus flavescens

All the members of the *Celeus* genus sport impressive crests, but the Blond-crested has a spikey hairdo that is outrageous. This stunning woodpecker is typically found in rainforest and well-wooded savanna in Brazil, Argentina and Paraguay, but, in some areas, it can also frequent plantations, orchards and gardens too. It eagerly eats fruit and seeds but mainly feeds on tree-dwelling ants and termites – excavating its nesting hole in their arboreal nests.

Black-headed Woodpecker
Picus erythropygius

This handsome woodpecker, a relative of our Green Woodpecker, is confined to dry subtropical forests, chiefly of mature dipterocarp trees, in parts of Thailand, Myanmar, Laos, Cambodia and Vietnam. It occurs over a large area, but is often scarce within it, with populations fragmented and isolated. A social species, noisy family groups search together for insects at all levels from the forest canopy down to the ground and are often joined by other birds in mixed feeding flocks.

Bearded Woodpecker
Dendropicos namaquus

At 24-26cm (9.4–10.2in) in length, this is one of the largest woodpeckers in Africa. It ranges from Sudan and Eritrea in the north to South Africa, and is quite common in dry wooded savanna. If you are ever on safari, look out for it in game camps. Both sexes are noisy and raucous in the period before nesting, making rapid, rattling, chattering calls. Both drum, too, producing distinctive slow rolls of around a dozen beats that seem to hesitate and stutter before ending in a few solid knocks.

Woodpeckers that avoid wood

As a group, woodpeckers have adapted to live in all kinds wooded habitats, but some species are found in places where there are few trees. Strange as it seems there are those that hardly ever visit trees and rarely peck wood, with some foraging and nesting in cacti in deserts and others in bamboo thickets. Many will drop to the ground to search for food (as mentioned earlier, the Green Woodpecker does this the most often in Britain) and several routinely feed in grasslands, but these species still excavate their nesting and roosting holes in trees. Two species have taken things a step further, spending almost all their time on terra firma: the Andean Flicker (*Colaptes rupicola*) of the high Andes in South America, and the appropriately named Ground Woodpecker (*Geocolaptes olivaceus*) that lives in the mountains of South Africa and Lesotho. These two woodpeckers live at or above the tree-line (where trees end and open country begins, high up in the mountains) and are the most terrestrial on the planet, walking and hopping rather than climbing, digging into soil and probing turf and tussocks for insect prey. They also excavate their holes (burrows or tunnels) in the ground, earth banks, road cuttings or between rocks rather than in trees. The Andean Flicker also makes communal roosting burrows in the walls of mud buildings.

Above: Andean Flickers live in largely treeless terrain at high elevations in Argentina, Bolivia, Chile, Ecuador and Peru.

Below: Ground Woodpeckers rarely perch in trees; in fact there are often no trees at all where they live! These birds are on a boulder high in the Sani Pass in Lesotho.

Social not sociable

Most species of woodpeckers are loners. They lead mostly solitary lives, seldom gathering in flocks, roosting together, or foraging close to one another. Most are only social when it is time to breed, but even then, it often seems that male and female woodpeckers just tolerate each other while they raise a brood. Once they have mated, woodpecker parents do most things (excavation, incubation, brooding and feeding of the chicks) in parallel, not together. There are times, of course, when woodpeckers can be seen in family parties, but as a rule, individuality is the way. Britain's three resident woodpeckers usually conform to the above, but elsewhere in the woodpecker world things are not always so. Wrynecks are often seen in pairs and are sometimes found in groups when on migration. Piculets often go about their business in pairs, whereas in Asia, Great Slaty Woodpeckers live in extended families, as do Magellanic Woodpeckers (*Campephilus magellanicus*) in South America,

and Lewis's Woodpeckers (*Melanerpes lewis*) migrate in flocks. Indeed, the Americas are home to some gregarious species. Two of the most studied are the Red-cockaded Woodpecker (*Picoides borealis*) and Acorn Woodpecker (*Melanerpes formicivorus*), both of which are highly communal and employ what scientists call a 'cooperative breeding system'. They live in groups called clans, typically made up of four to five individuals, including one or more breeding pairs and non-breeding 'helpers', usually young from previous years, which assist in feeding the brood and defending the territory from intruders. Clan dynamics are complex, with polygamous mating patterns observed, such as one male with several females (*polygyny*), one female with several males (*polyandry*) and paternity not always clear, at least to human observers. All very different from the traditional monogamy that the British woodpeckers tend to practice!

Magellanic Woodpecker

Acorn Woodpecker

Migrating woodpeckers

Migration to and from breeding areas is rare in woodpeckers. Most species don't make regular seasonal movements as, say, swallows do. Some will wander, especially in winter, in response to a lack of food or harsh weather, but perhaps 95 per cent of species are sedentary, meaning they stay in the same area all year round. Of the British woodpeckers, only the Wryneck is truly migratory; the others are resident. Indeed the Green and Lesser Spotted are highly sedentary, seldom straying from the area in which they fledged. Great Spotted Woodpeckers are more adventurous: with young birds dispersing once their parents finally refuse to tolerate them, continental populations crossing the North Sea to Britain when conifer seed crops there fail, and birds from Britain pioneering the colonisation of Ireland. There is a case of a female 'migrating' back and forth between Hertfordshire and Northamptonshire (about 99km or 62 miles), but ringing recoveries of Great Spotteds in England have shown that most never move more than 40km (25 miles) from where they were ringed.

Above: Wrynecks are one of the very few woodpeckers that migrate. This Wryneck was caught and ringed in the autumn as it moved between its winter and summer homes.

Above: This female Yellow-bellied Sapsucker spent most of her winter in coconut palms by the 'Bay of Pigs' in Cuba, having fled the cold of Canada.

Elsewhere, there are woodpeckers which behave differently. In autumn, Rufous-bellied Woodpeckers (*Dendrocopos hyperythrus*), an eastern cousin of our Great Spotted, migrate from north-east China and south-east Russia to winter in southern China. Several Northern American woodpeckers, including Lewis's Woodpecker and the Northern Flicker (*Colaptes auratus*), also move seasonally between the cold north and the warmer climes of the southern states. The longest journey of all woodpeckers is undertaken by the Yellow-bellied Sapsuckers (*Sphyrapicus varius*) that leave the frozen forests of Canada every autumn to enjoy the winter sun of Central America and the Caribbean.

Endangered or extinct?

At up to 60cm (23.6in) long, the Imperial Woodpecker (*Campephilus imperialis*) is almost certainly the largest woodpecker to have ever existed. This giant once lived in the Sierra Madre Occidental mountains in Mexico, but sadly has not been seen with certainty since 1956. Logging of its pine forest habitat, hunting by native peoples for food and the trade in body parts for imaginary medicinal properties and lucky charms were exacerbated by the collecting of this increasingly rare species for museums and private collections, and all of these factors combined to bring about its demise… or did they? The Imperial Woodpecker has not been officially declared extinct – it is listed as *Critically Endangered: Possibly Extinct*. In recent years expeditions to locate the bird have been conducted, but nothing found, and now only the very hopeful believe that this magnificent woodpecker may persist, hidden in the most remote and inaccessible forests.

In 2005 some sensational news hit the headlines. It was announced that the Ivory-billed Woodpecker (*Campephilus principalis*), a bird not seen in the USA for sure since 1944, had been spotted in Arkansas, in April 2004. A close relative of the Imperial, the Ivory-bill was once found throughout the vast tracts of natural woodlands of south-east USA and Cuba, but a familiar tale of the overexploitation of timber, hunting and collecting proved calamitous and the species was presumed extinct. The evidence from the 'rediscovery' consisted of brief sightings, vague video footage and inconclusive sound recordings, but unfortunately no photographs. Some were not persuaded that the bird had been rediscovered. Several expeditions to find this enigmatic woodpecker, in both the south-eastern USA and in Cuba where it was last seen in 1987, have since been made but still no hard evidence has been obtained. Tantalisingly, the Ivory-bill, like the Imperial, has not been declared extinct and is still classified as *Critically Endangered*.

Above: Tragically, the only Imperial Woodpecker you are likely to see will not be in Mexico, but a stuffed specimen like this one in Vienna's Natural History Museum.

Left: Though some are still searching the wooded swamps of the south-eastern USA, it seems the iconic Ivory-billed Woodpecker has followed the Passenger Pigeon and Dodo into extinction.

Anatomy and Adaptations

Woodpeckers are one of the most specialised bird families. They can land vertically on the side of a tree trunk and stay there with ease, seemingly defying gravity, and knock on and bore into trees repeatedly without getting injured. If we were to continually hit our heads on hard surfaces, as woodpeckers do many times every day when foraging or drumming, we would die – woodpeckers can bear G-forces of over 1,200g while humans would be concussed at a G-force of less than 100g. How are woodpeckers able to do this? Well, it's all down to the unusual anatomy of these avian headbangers.

Carpenters

Above: The great naturalist and scientist Charles Darwin commented on and was impressed by the wonderfully evolved anatomy of woodpeckers.

Most woodpeckers spend their time in trees where they hack and peck to create nesting holes and get at hidden insect food that most other birds cannot reach. Every part of their anatomy has evolved to make these jobs easier. Charles Darwin remarked that the woodpecker's *'feet, tail, beak and tongue… are so admirably adapted to catch insects under the bark of trees'*. In evolutionary terms, their anatomy is a winner. It's fantastically functional, each part playing a role and combining to make them nature's master carpenters. Let's look at the physical attributes that woodpeckers possess and which so impressed the great scientist.

Above: Woodpeckers pound on trees at speeds that would leave humans with a very severe headache… and probably worse.

Opposite: A female Great Spotted Woodpecker rests between bouts of hacking holes in a rotten tree trunk.

Bill The woodpecker's bill is a multipurpose tool that is used to hammer, split, lever, tap and probe wood. It is made of hard but flexible bone, is straight, broad-based, chisel-tipped and self-sharpening. They vary in design to suit their purpose: the Great Spotted Woodpecker, which frequently works wood and drums regularly, has a broader and flatter bill, while the Green Woodpecker, which forages on the ground and drums less, has a longer and more decurved (turned downwards) one. Just how important the bill is can be seen by looking at nestlings – even before they have grown any proper feathers their bills are large and developed. Furthermore, the bill's axis is set very low on the skull which means that potentially damaging impact shocks are deflected below the brain.

Below: The woodpecker bill is an impressive and versatile tool. Most birds use it to work on wood and some, like the Green Woodpecker, use it to dig into soil and probe turf for ants.

Nostril bristles Many birds, particularly insect-eaters (including woodpeckers) have tufts of hair-like feathers over their nostrils (called rictal bristles), but their exact purpose is still not fully understood. It may be that they protect the eyes from debris when in flight, or help in catching insect prey or funnel it towards the open beak – fine for nightjars and flycatchers, but not so in the case of woodpeckers. Perhaps nature's carpenters have developed their very own dust-masks just like human woodworkers have.

Above: Rictal bristles are hairy tufts at the base of the bill that may act as a filter, helping stop dust and splinters of wood from getting into a woodpecker's eyes and up its nose. They are just one of the many bodily adaptions these amazing birds have evolved.

Maxilla This plate of spongy tissue lies between the upper mandible and the skull and functions as a 'shock absorber', spreading impact forces away from the brain. The maxilla is unique to woodpeckers and more developed in arboreal species, like Great Spotted and Lesser Spotted Woodpeckers, than in those that spend less time hammering on trees, like Green Woodpeckers.

Skull The woodpecker skull is thick, made of supple bone and positioned above the line of the bill. Compared to most other birds, the gap between the skull and the brain is narrow and filled with relatively little cerebral fluid so the brain does not bounce around. All these features combine to create a 'buffer zone' that dilutes and diverts potentially harmful vibrations and shocks.

Skeleton All birds have the same basic bone structure, although it is shaped very differently according to their lifestyles. Woodpeckers have light skeletons like all flying birds, but with some fascinating adaptations. The ribs are broader than in similar-sized birds with the second pair, which support the shoulder blades (scapulae), being particularly thick and sturdy – this makes a solid base for hard-working neck muscles. The last bone in the vertebral column, called the ploughshare (the pygostyle) is proportionally bigger and flatter than in all other birds and affords firm support for the tail feather 'props'. The bony plate between the eye sockets is thicker in woodpeckers than in most other birds and provides

Below: Woodpeckers have a light but strong skeleton that means they can fly with ease, but at the same time cope with a demanding life, climbing up and down trees and banging on and hacking into timber.

a sturdy base for the tendons and muscles so they do not get unhinged during heavy hitting work. Most birds have a deep ridge along the breastbone (the sternum) called the keel to which the flight muscles are fixed. Woodpeckers have a shallow keel which allows them to be the good tree huggers they are.

Tongue Around 1490 Leonardo da Vinci made a note to himself that he should *'describe the tongue of the woodpecker'*. There are some amazing tongues in the bird world, but that brilliant man knew that the woodpecker's was really something. They are long and retractable and fall into two basic types. The first type is rigid and heavily barbed like a harpoon; the second is long and supple for lapping up prey like a lasso. The harpoon-like tongue is sticky and covered with many sharp barbs which angle backwards away from the tip. Prey is not impaled but rather hooked and glued, with no chance of escape. Woodpeckers are not the only birds to have barbed tongues; for example, penguins have them so that slippery fish suppers don't get out of the beak. The

Above: The Italian Renaissance genius Leonardo da Vinci, a prolific mathematician, inventor and artist, was intrigued by the incredible woodpecker tongue.

Below: Woodpeckers have barbed tongues – once hooked, insects have no chance of escape. This is the tongue of a Green Woodpecker.

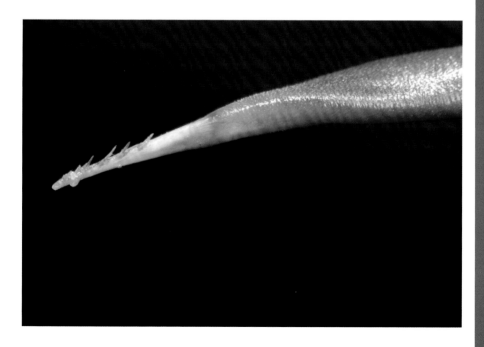

Right: This Wryneck has stuck out its long, sticky tongue and is using it to catch ants on the ground.

lasso-like tongue is very long, flattish, sticky and rich in sensors at the tip. Using two special muscles at the base, woodpeckers with this kind of tongue can move it inside and around insect galleries (burrows). They extend their tongues way out of their bills to catch prey; just how far depends on the species – Green Woodpeckers can poke theirs 10cm (3.9in) beyond the tip of their bill.

Hyoid horns All vertebrates (animals with backbones) have a set of bones, sinew and cartilage in the neck called the hyoid which supports the muscles that control the tongue. In humans, it is a rather small U-shaped structure; in most birds it is more complex with two extensions, called horns, that extend to the tip of the tongue. In woodpeckers, the hyoid horns are very long and when moved forward push the tongue right out of the bill where it becomes the nemesis of insect prey. When the tongue is pulled back in, the hyoid is wrapped and stored around the skull and, ingeniously, because of its elasticity, it adds a further protective cushion to the brain.

Below: The hyoid horn apparatus (shown in pink) – another weird and wonderful woodpecker anatomical adaptation.

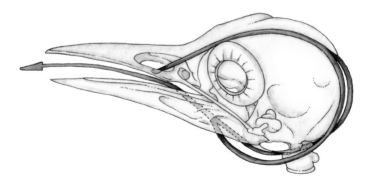

Salivary glands These are bigger than in birds of a similar size and copiously coat the tongue with a mucus in which prey gets glued. A gland at the base of the skull also secretes a sticky drool that is thought to catch any wood dust that enters the head when a bird is excavating.

Limbs The foot arrangements for all birds vary to suit their purpose, and woodpeckers need sturdy legs and strong claws for their way of life. Perching birds have three toes pointing forward and one pointing back (the hallux), which is the innermost digit and equivalent to our big toe. This arrangement is called anisodactyl. Other birds, like cuckoos, parrots and barbets, have two toes pointing forward and two pointing back in a formation called zygodactyl. This form is not ideal for climbing, but it does suit some woodpecker species that spend more time on the ground, like Green Woodpeckers. Truly arboreal species have developed the arrangement further with the hallux placed to the side and hardly ever used (ectropodactyl) or elongated and splayed, or even missing totally as in the three-toed species. In the case of large woodpeckers, a four-toed, splayed formation may help in weight bearing and stability, but why a four-toed and three-toed variation developed is not fully understood. All British species have four toes on each foot.

Below: Most woodpecker species have four toes on each foot. Note the rough and rugged skin and hooked, crampon-like claws – ideal gripping and climbing gear.

How many toes?

The fittingly named Eurasian Three-toed Woodpecker (*Picoides tridactylus*) and American Three-toed Woodpecker (*Picoides dorsalis*), as the well as the Bamboo Woodpecker (*Gecinulus viridis*), Common Flameback (*Dinopium javanense*), Rufous Piculet (*Sasia abnormis*), White-browed Piculet (*Sasia ochracea*) and several others in Asia, all have three toes (tridactyl). In all cases the missing toe is the small hallux.

Eyes Birds, fish, amphibians and reptiles keep their eyes lubricated and free of foreign bodies not by blinking with two eyelids as we do, but by using a third called the nictitating membrane. These natural 'windscreen wipers', which are thicker in woodpeckers than in most other birds, shut automatically a millisecond before a woodpecker hits a tree with its bill and prevent dust and debris getting in the eyes.

Right: A female Great Spotted Woodpecker protects her eyes from flying splinters by closing her nictitating membrane.

Ears The sensitive inner ear has an extra thick membrane and there is also a strengthened bone in the middle ear (the columella) which protects it from damaging vibrations and shockwaves and probably also from the din of drumming, pecking and hammering.

Skin Any taxidermist who has prepared a specimen (perhaps for a museum) will tell you that woodpecker skin is tough and firmly fixed to the body and skull – more so than on songbirds. Robust skin is good protection against wood splinters and the chemical sprays and bites of ants.

Feathers Feathers are unique to birds although dinosaurs, their ancient ancestors, did develop them. They are mostly made of a tough protein called keratin, which gives strength and flexibility, and from which hair, horns, hooves, beaks and fingernails are also made. Once grown it is essentially dead with no nerves or feeling. Woodpeckers come in many colours but most have black tail feathers caused by a pigment, melanin, which also adds strength and wear resistance. These qualities are needed by woodpeckers for going about their daily lives. All three of Britain's true woodpeckers have black central tail feathers with stiff shafts (the rachis). Strong tails act as props to keep woodpeckers on vertical tree surfaces and are long to place the centre of gravity nearer to the trunk.

Above: Woodpeckers have particularly strong and sturdy tail feathers that provide support, acting like props, when they work on trees.

Bionic birds?

This adapted woodpecker body means they can easily outdo humans in G-force tolerance. Given that fighter pilots and astronauts need special 'anti-g suits' to avoid blacking out when encountering G-forces that woodpeckers experience daily, we might well think they are bionic. Obviously, the reality is woodpeckers are far from indestructible. Woodpeckers are uniquely designed to cope with high g-decelerations related to the stresses and shocks to their brains and bodies from the things they do. Drumming and pecking on trees is one thing; flying into a window or being hit by a car, as happens with urban-living woodpeckers, is quite another and they do not survive such impacts.

Below: Some woodpeckers have adapted to live in urban environments, but such places can be hazardous. Despite evolving impressive impact-tolerating bodies, they rarely survive collisions with windows and vehicles. Here is an unfortunate Great Spotted Woodpecker.

Behaviour

All woodpeckers are active during the day, but they are seldom around at dawn, unlike most songbirds. They like a lie in, and an early night too, usually going into their roost hole well before dark. Yet once up and about woodpeckers are doers, cramming their day with chores – finding food, making a hole, and letting their rivals and partners know they are around.

Everyone has heard the tongue-twister *'How much wood would a woodpecker peck, if a woodpecker could peck wood?'* Well the answer is probably an awful lot! Once one finds a tree crawling with insects, it will often work there incessantly, hacking away, flicking aside wood chips, digging deep to exploit the bug banquet to the full. Quite simply, woodpeckers seem to have an insatiable need to use their bill – they are born to peck.

Opposite: Hanging about – by using its strong claws and stiff tail a Green Woodpecker seemingly defies gravity.

Below: Think you're good at tongue-twisters? This Wryneck shows us all how it's really done!

Friend or foe?

There are two male Green Woodpeckers on the lawn, staring at each other. One flicks its long tongue in and out. Then they point their bills skywards and one draws a figure-of-eight pattern in the air. They then lower their heads and touch bills like two fencers before a bout. Then it is *allez* and they lunge and joust, bobbing and swaying their heads from side to side and flicking wings. They chatter, then suddenly stop, freeze and point their bills at each other, in an apparent stand-off. It all seems staged and synchronised.

Above: Great Spotted Woodpeckers in a feisty food fight at a peanut feeder.

In a similar scenario, two Great Spotted Woodpeckers chase each other noisily around a tree trunk a few feet off the ground. Then they stop, on opposite sides of the tree, as if nothing has happened. After what seems like an age, they start again, ruffling their crown feathers, flicking wings, spreading tails, all the while calling excitedly. Testing each other out like two stags. Then they pause once more. One of the birds starts to lightly peck the tree as if he has suddenly become hungry and needs to find food – this behaviour is called displacement activity. On another day, he might have done a little drumming.

Fights between woodpeckers can be noisy affairs or silent ones where it seems a case of 'let's see who blinks first' prevails. Bodily contact, where birds get injured, is rare although the bill is a potentially lethal weapon when jabbed at an adversary. Things get physical in disputes over holes or foraging places like an ant hill or a garden peanut feeder but generally, squabbles rather than full-on fights are the norm. Disputes are typical when two testosterone-fuelled males meet, but the behaviour of a male towards a female is often similar. For example, a male might jab his bill towards his mate and the bills may touch, but this may also be a form of symbolic feeding. It is often hard to tell with woodpeckers as their antagonistic and courtship displays are alike.

Courtship

Woodpeckers are ready to breed at one year old. The courtship period is short, and compared to many other birds the rituals are simple. Almost everything takes place near to the future nest hole. The first thing males do to attract a mate is drum. Once a female shows interest and begins to hang around, the male will perform fluttering flights and tap at the nest hole. When she is suitably impressed, it is the female that usually initiates

mating by squatting low on a bough near the nest to attract her mate. If a rival appears at any time and is seen off, the paired male will take the initiative and fly to the female to copulate. After mating males usually fly away or go to the hole, and sometimes both sexes start to preen. Despite what often seems like an uneasy relationship, woodpecker pairs are usually monogamous and faithful to each other with extra-pair mating rare.

Above: A pair of Green Woodpeckers mating. Woodpeckers invariably mate close to their future nesting hole, often high up on a horizontal branch.

Below: A pair of Lesser Spotted Woodpeckers mating. Mating is usually brief and often involves a flurry of wings.

Dealings with others

Above: A Great Spotted Woodpecker and a Great Tit, two common woodland and garden birds, eye each other up.

Below: When food is plentiful birds will often flock into gardens in great numbers. On the right, a juvenile Great Spotted Woodpecker shares a peanut feeder with a Great Tit.

Interactions between animals can be divided into two basic kinds: *interspecific* and *intraspecific*. For our purposes, we can define these terms simply as woodpeckers interacting with another species (interspecific), and woodpeckers interacting with one of their own species (intraspecific).

One of the common ways animals interact interspecifically with one another is in a predator–prey relationship – one wants to eat the other, and one wants to avoid being eaten. Woodpeckers are sometimes on the menu, but not that often – none of the British woodpeckers are the main prey of any predator. Ants are not so lucky. They are the favourite food of both Green Woodpeckers and Wrynecks. Indeed, food is often an issue with woodpeckers as they sometimes want to eat what others are eating and vice versa. Most people who feed birds in their garden will have seen how a pecking order operates: it is often the biggest bird that is first at the table, or the most aggressive, and in suburban Britain Great Spotted Woodpeckers tick both boxes. In

a confrontation over suet or seeds, tits and finches don't stand a chance. Besides food, the other area where woodpeckers and other animals frequently disagree is over holes. After all, it must be infuriating to have squatters like Starlings trying to move into your new home after you have spent weeks building it.

In Britain, Green Woodpeckers don't often get into any dealings with their smaller cousins as they seldom meet, tending to hang out in different places to Great Spotted and Lesser Spotted Woodpeckers, and Wrynecks are just too thin on the ground. However, the two pied woodpeckers do interact – Great Spotteds will interfere with nesting Lesser Spotteds, sometimes even predating their young, and it is sometimes said that this is one of the main reasons why the smaller woodpecker has such a low rate of breeding success in Britain. However, there is little evidence to support this view, and they coexist well in the rest of Europe. Woodpeckers are known to prey on other woodland birds, too, but studies examining their impact have shown that they do not cause significant declines in the populations of any.

Above: Woodpeckers will occasionally predate the eggs and chicks of other hole-nesting birds. Here a Pied Flycatcher takes exception to the presence of a Great Spotted Woodpecker.

BEHAVIOUR

The woodpecker's new clothes

Above: Woodpecker feathers are delicate and flexible yet strong and durable. Here are two secondary flight feathers (which run along the 'arm' of the wing): on the left, a Great Spotted Woodpecker's; on the right, a Green Woodpecker's.

Feathers may seem delicate but they are actually quite tough. Woodpecker feathers are some of the most durable in the bird world, and they need to be as they are rubbed against rough tree bark every day. Despite birds doing their best to keep their feathers in good condition by preening and bathing, they do not last forever. Feathers become worn, faded and are under constant attack from nibbling mites and lice, and, as they do not grow continuously, as fur and hair does, they need to be replaced.

The process of shedding of old feathers and replacing them with new ones is called moulting. Woodpeckers have 10 primaries (main flight feathers), 10 to 12 secondaries (supporting flight feathers), and 12 rectrices (tail feathers) although the outermost pair are short, soft and sometimes barely noticeable. The order in which adults replace their feathers (the moult regime) is coordinated and staggered so that they can continue their daily activities without being unduly hindered. The true woodpeckers only have one complete moult a year, just after breeding is over, and the flight feathers are changed in sequence, not all at once, so that they are always able to fly. The two central tail feathers, which are so important as props when woodpeckers cling to vertical tree trunks (see page 39), are changed last so that during the overall moult the birds can still carry on working. Woodpeckers are often very quiet when moulting as they're most vulnerable.

Anting

No one is quite sure what anting is, but it is generally thought that it is a form of behaviour where birds rid their feathers of mites and other parasites with the formic acid that ants emit when disturbed. Another idea is that anting is food preparation, where ants are encouraged to eject their acid before being eaten in a tastier form. More controversially, some have suggested that anting is simply comfort behaviour, like sunning or dusting – birds do it because it feels good to be stimulated by sprayed acid. Anyway, there are two kinds of anting: passive and active. The first involves the bird simply sitting amongst ants which swarm over them spraying their chemicals. Birds using this method spread their wings and tail, lie back and let it happen. In active anting, the ants are picked up in the bill and dabbed directly onto the feathers. Anting is probably a quite common bird behaviour, but seldom observed. Green Woodpeckers, which spend most of their lives around ants, certainly do it and, whatever function it serves, probably feel better for it.

Above: A Green Woodpecker lies down, spreads its wings and indulges in a bout of anting.

Mimicking snakes and playing possum

Above: Wrynecks do not really look like 'typical' woodpeckers and indeed often take up some decidedly uncharacteristic poses.

Wrynecks can, as their name suggests, twist their neck awry. A bird will raise its crown feathers, open its bill and slowly writhe and stretch out its neck before suddenly recoiling it, and some may also make hissing alarm calls as they do so. It is extraordinary behaviour which both adults and young perform when they feel threatened – when a nest box is being inspected, for example. When handled by a ringer, a bird will sometimes rotate its neck up to 180 degrees, but others do not act out the whole show, only swaying their head from side to side, and occasionally they will 'play possum', hanging limp with their eyes shut, pretending to be dead. Predictably, the Wryneck's writhing performance has been compared to that of a

snake defending itself from a predator, and in the past this led to all sorts of superstitions surrounding the bird, usually unfavourable. Currently the most common explanation for this bizarre behaviour is that it is mimicry, a direct imitation by the bird of the movements of a snake when threatened, but tempting as this explanation is, no one has yet been able to offer firm evidence to prove it.

Above: True to their name, Wrynecks do twist their necks awry. This head-bending, swaying display is mostly witnessed when they are handled by ringers.

The wise woodpecker

We have all heard the phrase 'bird brained', but as it happens, birds are not dim at all. When it comes to intelligence in birds, parrots and crows are generally regarded as the 'brainy' families – some parrots 'talk' and you may have seen those TV documentaries where crows use tools. It's true that in terms of relative size, parrots and crows do have bigger than average brains, but so do owls, hornbills and woodpeckers. Research at the world-renowned Konrad Lorenz Institute in Vienna, for example, has shown that the Great Spotted Woodpecker has about 1.6 times more volume of grey matter than the similarly sized Blackbird (*Turdus merula*).

Researchers regard extracting hidden food (as woodpeckers do when they forage and feed) as a sign of intelligence as it needs more brainpower than simply picking up things. Much of their insect prey lies out of sight in timber and, though they may hear or 'feel' it, they show judgement and inventiveness in getting at it.

Below: When foraging, Woodpeckers often show resourcefulness and ingenuity, which are regarded as signs of intelligence. Here a young Green Woodpecker probes the ground, using its bill like a spade, to unearth hidden prey.

Studies with captive woodpeckers have shown that when food is the reward they are equal to some mammals in completing complex tasks. With minimal 'training' they quickly learn to remove corks to access hidden food items, can associate different colours with different kinds of food and can match acorns and nuts to the right sized holes. In a remarkable study, twelve-week-old Great Spotted Woodpeckers learned different drumming codes for different kinds of food – one strike for a pistachio nut, two for a house cricket, three for a mealworm, and so on.

Anvil use (see page 64) is another area where woodpeckers show how brainy they are. Woodpeckers carry hard or oversized food items to cracks or crevices in trees or walls where they wedge and process them. Some species take it further, modifying their anvils, cleaning, enlarging and shaping them to suit and, in some cases, they are used for years and become regular woodpecker workshops. Anvils are tools (or at least proto-tools) and tool use by animals is looked upon by behavioural scientists as a sign of intelligence.

Above: A pile of 'processed' conifer cones below a regularly used woodpecker 'workshop' or 'anvil'.

Below: This tree stump in a forest in Estonia has been 'beaten up' by a Black Woodpecker in search of insect food.

Food and Foraging

Most of a woodpecker's day is spent foraging. Some are food specialists, others are generalists, but all can be opportunistic – ready to take advantage of unexpected abundances in resources. Despite all their biological adaptations and foraging skills, like most animals, woodpeckers seldom turn up their beaks at an easy meal.

As far as food goes, there are three basic kinds of woodpecker: insectivores (insect-eaters), frugivores (fruit and nut-eaters) and omnivores (not that fussy). Most, including those in Britain, are primarily eaters of insects and other invertebrates: beetles, woodlice, flies, leatherjackets, caterpillars, bee and wasp grubs, spiders, snails and ants are all eaten. Ants are the preferred prey of many, with some feeding almost entirely upon them. Let's look in more detail at the diets of our British species.

Green Woodpeckers have the narrowest diet of all European species. They eat insects of all kinds, snails and earthworms, and are not against berries and the odd piece of windfall fruit, but ultimately, they are ant-eaters. Studies have shown that when these insects are plentiful, they feast on little else, taking them in all stages – adults, pupae, larvae, eggs. Once a productive colony is located, it is returned to repeatedly and it is no exaggeration to say that a hungry brood of chicks can be fed millions of ants in one season. Different ants are taken seasonally, and

Opposite: A male Lesser Spotted Woodpecker with a bill full of food.

Above: A young Great Spotted Woodpecker with a cherry.

Far left: When it comes to food, most woodpeckers are not that fussy. Woodlice are just one of the species they find tasty.

Left: An adult male Great Spotted Woodpecker with a nut.

Above: A male Green Woodpecker in a classic pose – sitting on grassy ground looking for ants.

Below: The ground-dwelling Yellow Meadow Ant is one of the favourite foods of Green Woodpeckers in Britain.

local availability obviously influences the menu, but they are experts in feeding upon those that live in open areas such as *Formica, Myrmica* and *Lasius* species – in Britain the Yellow Meadow Ant (*Lasius flavus*) is a favourite. Birds avoid dense forests, foraging basically anywhere where the grass is short – sheep-grazed pastures, woodland glades, lawns – but will visit orchards and young plantations. The presence of ants is the main factor in the distribution of the Green Woodpecker and seeing as most ants in Britain are found in the southern counties (where the highest summer soil temperatures are) it's not surprising that the Green Woodpecker is most common there.

When it comes to food, Great Spotted Woodpeckers are not as choosy as their larger cousin. They are more omnivorous and resourceful, eating all kinds of insects, pine seeds, acorns, nuts, fruit, berries and sap, and this adaptability is one of the keys to their success. Their diet varies locally and changes with the seasons – in spring and summer insects dominate the menu, but in autumn and winter they turn to more vegetarian fare. They of course have all the tools and skills of a typical woodpecker and are expert in extracting insects from inside timber, but nevertheless they readily take other food when it is on offer. That very British habit of feeding garden birds, for example, has benefitted Great Spotted Woodpeckers, and they soon get used to visiting bird tables and feeders, often favouring suet, peanuts and sunflower seeds. They can also, somewhat notoriously, take a liking to the nestlings of cavity-nesting birds such as tits, especially from nest boxes, and have been known to enlarge the entrance hole to get in. Great Spotted Woodpeckers routinely do this once they realise that chicks are plump, rich in protein and much easier to obtain than beetle larvae burrowed deep within trees. In attempts at preventing this kind of thing, nest boxes with protective metal plates around the entrance hole are available, although some woodpeckers, showing characteristic inventiveness, simply hack a hole elsewhere on the box.

Above: A Great Spotted Woodpecker extracting seeds from a conifer cone.

Below: An acrobatic Great Spotted Woodpecker hangs below its lunch of peanuts.

Above: A female Lesser Spotted Woodpecker pecking into rotting wood in search of an insect meal.

Lesser Spotted Woodpeckers are highly insectivorous, although they do occasionally eat some fruit and veg, especially berries. Being small they specialise in prey such as aphids, ants, caterpillars, spiders, woodlice and smaller bark beetles, leaving the bigger wood-boring ones that live deep in wood to their cousin the Great Spotted Woodpecker. In winter, they feed on moth larvae hiding in catkins, which are likely to be an important food source in the lead up to the breeding season, as are beetle larvae found in softer, moist timber. Then in spring, they switch to predominantly defoliating leaf caterpillars, sawfly larvae and the like to feed their young. They will visit garden feeders, to take suet and sunflower seeds, but as they are now so rare in Britain it's unlikely one will drop by – if you are lucky enough to get one on your feeders report it to the BTO! In spring and summer, they forage mostly high in the crowns of deciduous trees and on slender snags, gleaning surface-dwelling insects, and in winter they turn more to the beetles that hide under bark or in rotting wood.

Below: Wrynecks are avian ant-eaters, that habitually search for ground-dwelling ants in meadows and pastures.

Wrynecks are ant-eaters, feeding almost totally on the larvae and pupae of small, terrestrial species such as the genus *Lasius*. Prey is mostly lapped up directly from bare ground with their long gluey tongue, and birds return repeatedly to profitable spots. Naturally, some other food is eaten, especially small insects and snails encountered on the ground. Fascinatingly, birds have been seen giving their nestlings bits of snail shells. As mentioned earlier Wrynecks do not have the bodily tools to hack into trees to get at beetles and the like.

Fruit and nut cases

All woodpeckers eat insects, but there are several that are also frugivorous, consuming a good deal of fruit, berries, seeds, nuts and nectar. These are mainly tropical woodpeckers, such as many in the *Melanerpes* and *Celeus* genera, which consequently play important roles as seed dispersers and pollinators. Sapsuckers, as you might guess, like sap. Great Spotted Woodpeckers eat more plant matter than the other British species, especially pine and spruce seeds, and willingly accept free vegetarian meals from garden feeders. Very few woodpeckers have been recorded feeding their young fruit and nuts: for most, nourishing protein-rich invertebrates are the food of choice for growing nestlings.

Above: Although woodpeckers as a group are mainly meat-eaters, eating a variety of insects, many often add fruit and veg to their diets. This female Great Spotted Woodpecker has collected a nut.

How do woodpeckers find food?

How woodpeckers know where to find hidden insect prey is not fully understood, but it is clear they do not just hack away indiscriminately at any old tree; that would be an enormous waste of time and energy.

Sight Visual signs are certainly important. In experiments with Pileated Woodpeckers (*Dryocopus pileatus*) it was found that they use ultraviolet perception to differentiate between wood substrates, which suggests that they target their foraging to the most productive timber.

Touch Green Woodpeckers use their tongues to detect ants, not just to catch them. The tongue is sensitive at the tip, with many sensors that probably detect vibrations as well as actual prey.

Above: Pileated Woodpeckers are native to North America. Studies on their foraging behaviour have revealed they use ultraviolet perception to explore specific areas on trees for grubs rather than searching haphazardly.

Right: Woodpeckers use their tongues not only to lick up prey, but also to locate it. Here a young Green Woodpecker pokes its tactile tongue into a hole in the ground.

Sound It's also likely that woodpeckers can hear prey that is out of sight, deep inside trees. Research on Lesser Spotted Woodpeckers found that they often foraged on those branches with the greatest abundance of beetle grubs, despite the prey being hidden. Green Woodpeckers dig into ant hills that have no ants on the surface and through snow to get at colonies. Perhaps the high-frequency sounds produced by insects are detected by woodpeckers.

Smell Very little is known about the sense of smell (olfaction) in birds and how important it is for finding food. Recent research suggests that this sense may be more developed in birds than previously thought, but at present there is no firm evidence that it is acute and important in woodpeckers.

Above: Using its stiff tail as support, a Green Woodpecker delves into an earth bank with its heavy bill in search of a meal.

Foraging techniques

Right: The base of this fungi-riddled and dying tree has been hacked open by a hungry woodpecker in search of the insect larva hidden within.

Below: When a tree is infested by bark-boring beetles, woodpeckers do not need to dig deep, but simply strip and peel away bark to get a meal.

Many techniques are employed by woodpeckers when searching for food, depending who is doing it, where it is being done and what they are after. Bark is scaled or stripped off, crevices probed, holes bored, wood cambium chiselled, logs split, ant hills dug into, foliage gleaned, turf and soil probed, leaf-litter swept aside and surfaces licked. Woodpeckers are purpose-built for working wood to get a meal, but once located, how do they catch their prey? When examined up close most tongues look sharp at the tip; however, their prey is not impaled or speared but glued in saliva, hooked by its barbs and then pulled out of their burrows.

Green Woodpeckers spend more time on the ground searching for food than any other European woodpecker. They probe into turf with their bill and sweep aside leaf-litter. Another tactic they use is to prod into an ant hill to provoke the insects to swarm out, and then lick them up with their long sticky tongue. Great Spotted Woodpeckers can be quite acrobatic (as you may have seen on garden feeders) but they tend to be more 'woodpecker-like', sticking to trunks and branches. Being small, Lesser Spotted Woodpeckers are the most agile of the British species, nimbly hanging from thin twigs to pick off prey or gleaning it from leaves.

Meals on the ground

Some woodpeckers forage for their food on the ground, rather than in trees, probing into soil for worms and insect larvae, sweeping away leaf-litter to expose prey, or searching for terrestrial ants and termites. As already mentioned, the most terrestrial species in Britain are the Wryneck and the Green Woodpecker, both of which need open bare ground and unimproved short grass pastures and meadows with good numbers of ants. Elsewhere, most of the New World flickers, many *Campethera* species in Africa, and the Green Woodpecker's *Picus* relatives in Eurasia routinely feed on the ground. Differences in the foraging habitats of woodpeckers are reflected in their comparative physiques, with ground-feeding species having less robust tail feathers than their more arboreal foraging relatives, reflecting the extra time they spend on the floor. On the other hand, ground feeders usually have longer and more decurved bills which they use as ploughs and spades rather than chisels.

Above: Though they nest in trees, Green Woodpeckers spend most of their time looking for food on and in the ground.

Below: Wrynecks, too, are often observed hunting for ants on the ground. This one is probing a hole with its long tongue.

Above: Not only is Lewis's Woodpecker a stunningly good-looking member of the family, it's also one of the very best at catching insects in flight, too.

Flycatching woodpeckers

Very occasionally a Great Spotted or Lesser Spotted Woodpecker might try to impersonate a flycatcher, and attempt to take an insect in flight, but this is not something these species are very good at. Wrynecks also have been observed flycatching, but this is not the usual way they find food. Several of the New World *Melanerpes* genus, however, are expert flycatchers, with Lewis's Woodpecker of western North America arguably the master. A glistening bottle-green above and salmon-pink below, this handsome woodpecker perches patiently on snags and fence posts and then sallies up to snatch flying insects with apparent ease, before gliding elegantly back down to its perch. Several insects may be caught in one flight. It's an impressive sight and feat, given that it is by no means a small woodpecker at up to 29cm (11.4in) in length.

Sap Sucking

How much sap would a sapsucker sup? There are four species of sapsucker (genus *Sphyrapicus*), all found in North and Central America, but none of them suck tree sap, they sip it. Many other woodpeckers around the world eat sap, but sapsuckers are the real connoisseurs, drilling precise holes in trees to specifically tap this rich sugary food. These holes, called wells, are made in the sapwood – the soft layer of living tissue between the bark and the heartwood through which nutrients flow up and down trees. Hummingbirds, warblers, squirrels, mice, butterflies and others all visit sapsucker wells for free meals. Sapsuckers return to their wells repeatedly and sometimes get extra snacks in the form of insects that were drawn to the nutritious food, only to get stuck. They have been seen dunking insects into oozing sap and then feeding them to their young. None of the British species apparently drill sap wells, at least in Britain, although on mainland Europe, Great Spotted Woodpeckers do occasionally girdle tree trunks with holes to get sap.

Below left: A Yellow-bellied Sapsucker busy drilling new sap wells.

Below: Rows of old, dried out sapsucker wells.

Anvils

Some woodpeckers use 'anvils', places where hard items of food like fruit stones, nuts in shells, conifer cones and hard-bodied beetles are wedged and then opened. Most anvils are simple, a crevice in the bark of a tree or stump, or just a fork between two branches, and are used just once. Great Spotted Woodpeckers will, however, specifically create and modify anvils to suit different kinds and sizes of food. These made-to-measure workshops are used repeatedly, even for years, and birds may have several in their territories.

Below: A Great Spotted Woodpecker at its anvil with a wedged conifer cone.

When a woodpecker finds a hard nut or bulky cone that it cannot easily open, it flies with it in its bill to the anvil, jams it in tightly and then sets to work. If the anvil contains a previously worked item, it will place what it is carrying between its breast and the tree, hold it there and then with its now free bill flick the offending item out of the anvil. Hundreds, sometimes thousands, of discarded cones or nutshell fragments pile up below these places.

Anvil use is most developed in Great Spotted Woodpeckers, especially when they are feeding upon pine and spruce cones which they wedge and then pry open to get at the seeds within. Anvil use is not unique to woodpeckers – some crows use them in a similar way, and the places where Song Thrushes smash snail shells are also called anvils – but when Great Spotted Woodpeckers adapt their workshops they are undoubtedly showing intelligence and arguably tool use (see page 51).

Stores and granaries

There are several woodpeckers around the world that store food such as nuts and mast in crevices, cavities or purposely drilled holes in trees. No British woodpecker does this, but the Acorn Woodpecker, one of the real characters of the woodpecker world, has taken this behaviour to an extreme level. These social birds create 'granaries' of thousands of holes, each one excavated to hold one acorn which is tightly wedged in. An incredible 50,000 acorns have been estimated in the biggest granaries.

Below: An Acorn Woodpecker at a 'granary'. These stores are meticulously maintained and can contain thousands of acorns.

Above: Adult male (below) and juvenile (above) Great Spotted Woodpeckers. Some parents will continue to feed their offspring for several weeks after they have left the nest.

Sharing supper

You may have walked in a woodland and seen a Great Spotted Woodpecker, heard a Green Woodpecker and, if you were in an exceptional place in Britain (and lucky or very talented), also come upon a Lesser Spotted Woodpecker, with nuthatches, treecreepers and other birds no doubt around as well. In continental Europe, seven or eight species of woodpecker can live almost side by side, and in tropical rainforests there might be a dozen or so. How is this possible? How can all these birds, some very similar, seemingly doing identical things, looking for the same food, coexist? Surely, they must be getting in each other's way, competing with one another? Well, perhaps things are not what they seem. There are overlaps, but there isn't as much competition as we humans might think.

Ecologists use several terms to describe how this habitat sharing works. *Sympatry* is when different species live in the same habitat, *niche partitioning* is how similar species use an environment differently to enable them to coexist, and *resource partitioning* is how species use the resources in an environment to avoid competition. Conflicts over food are avoided by various, often subtle, things. Woodlands host a smorgasbord of things to eat and there is a choice, for the picky woodpeckers (specialists like Green) and not so fussy ones (generalists like Great Spotted). Quite simply, although sympatric species often eat the same food, they also feed on very different things. Size also matters, with the vital statistics of woodpecker bodies, bills and tongues dictating where they can forage – Lesser Spotted Woodpeckers are more comfortable in canopy foliage than Great Spotted Woodpeckers are, but the latter are more at home on tree trunks and Green Woodpeckers are far happier on a lawn than both of their relatives. How they get food differs between species – some glean more, some bore more. Niche and resource partitioning occur between species, but also often between males and females of the same species. Though it is hard to see, the sexes differ in size, weight and bill length – males are generally bigger than females and these differences in dimensions influence where they search for food and what they eat.

Below: A young male Green Woodpecker gets a meal (almost certainly of ants) from his father.

Holes as Homes

Woodpeckers are considered to be one of the most successful bird families. One of the reasons for this is that millions of years ago, early on in their development, they evolved the skills to create safe places in which to nest – in cavities in trees.

Shelter and security

Birds build nests of many kinds: cups, mounds, platforms, saucers, scrapes, spheres, hanging socks, burrows and cavities. The advantages of raising a family and sleeping in the relatively stable environment of a tree cavity are many – in fact, it's hard to think of a disadvantage. Studies have shown that birds that nest in holes have higher levels of breeding success than birds that nest in the open, and this certainly goes for woodpeckers, which often succeed in fledging most and sometimes all their chicks. Two simple words can help explain this: *shelter* and *security*. Shelter from bad weather – rain, wind and cold – and better security when it comes to predators – only good tree climbers are a real danger.

Opposite: A well-grown Great Spotted Woodpecker nestling looks out at the world from the safety of the nest hole.

Right: A female Green Woodpecker is greeted by her chicks as she arrives at the nest hole.

Right: A Gila Woodpecker arrives at its hole in a saguaro cactus, which it shares with a Great Horned Owl.

Below: The White-browed Piculet of South East Asia often excavates its nest hole in bamboo.

All the woodpeckers on the planet today breed in holes, but not always in tree holes they have excavated themselves. Some species, like the Gila Woodpecker (*Melanerpes uropygialis*) and Gilded Flicker (*Colaptes chrysoides*) in the USA and Central America, make their nesting holes in cacti. In South East Asia, the Bamboo Woodpecker and Pale-headed Woodpecker (*Gecinulus grantia*) not only forage in bamboo poles, but also hack their nest holes in them, as do several of the tiny piculets. Others, such as the Rufous Woodpecker (*Micropternus brachyurus*) bore nesting cavities inside arboreal ant nests. The Campo Flicker (*Colaptes campestris*) in South America will use a hole in a termite mound and, as we heard earlier (see page 25), the Ground Woodpecker and Andean Flicker burrow into earth banks and between rocks. Some woodpeckers even use electricity poles and fence posts. The three resident British species tend to stick to what woodpeckers are supposed to do, and make their nesting holes in trees. Wherever the nesting hole is, when your young are blind, bald and helpless, as all woodpecker nestlings are, a cosy chamber is a great place to call home.

Excavation

Every spring, woodpeckers make holes in trees, even if they already have an ideal hole from the previous year where they successfully raised a family and that they intend to use again – some studies suggest only 25 per cent of holes may be reused. Both males and females excavate, but do not work all day; most work is done in the morning. Holes are seldom created in one go; some are started then completely abandoned, others revisited and finished off in the same season, and sometimes a year later. Working every day, Great Spotted Woodpeckers take about 20 days to make a nest hole from start to finish, but Green Woodpeckers tend to take longer, up to a month, as they are not as good at carpentry as their more arboreal relative. Then again, much depends on how many hours a day the birds work and the condition of the timber – obviously hard wood requires more effort than soft wood. Experience doubtless helps, so age may also be a factor as will the fitness of the bird.

Above: A Lesser Spotted Woodpecker working on a new hole. These little birds usually choose to excavate in soft, rotting wood.

Above: When working on a new hole, woodpeckers simply toss wood chips and wood dust out of the entrance to fall down to the ground below.

When work begins, strips of bark and wood chips are simply flicked aside and pile up at the base of the tree. Later, when a bird is working inside a developing chamber, it sticks its neck out from time to time and tosses debris away with shakes of the head. This nonchalant behaviour is something of a puzzle. Woodpeckers are generally wary of being seen by their nest holes. Despite this, they draw attention to them by letting a pile of wood chips amass below the tree in full view of any passing predator.

Although conifers are used by British woodpeckers, broadleaved trees are generally preferred for nesting sites, but it is often a case of what is available locally. Living trees are usually selected by Green and Great Spotted Woodpeckers, with holes excavated in an area of rotten wood. In England, favourite nesting trees for Green Woodpecker include ash, birch, oak, sweet chestnut and walnut. Lesser Spotted Woodpeckers usually nest in dead boughs or snags, particularly in softer species like alder, birch, poplar and willow, but also beech, elm and sycamore.

Orientation

The orientation of the entrance hole depends upon several things, but a suitable temperature in the cavity is vital. Hence in Europe, entrance holes tend to face southwards to benefit from natural solar heating and avoid cold winds. On hillsides, hole entrances usually face down the slope and small woodpeckers often position their holes on a leaning trunk or the underside of a branch to help shelter it from rain.

Cavities and chambers

A typical tree cavity consists of an entrance hole that leads to a vertical cavity, with a wider chamber at the bottom which can be used for nesting or roosting. Nesting cavities double as roosts outside the breeding season, but those cavities made just for overnight roosting are often more basic and not used for raising a brood. Woodpeckers do not take anything from outside into their nesting cavities to line it. There are no proper nesting materials; however, wood chips and dust from the excavation of the chamber wall help cushion the eggs on the hard wood floor. Camera footage inside cavities has shown adults pecking the chamber walls, but it is unclear whether this is done to add chips to the floor or to widen the chamber for the growing chicks.

Who made that?

You can guess which woodpecker made any hole you find in Britain by looking at the size of the entrance. Woodpeckers like to squeeze into their cavity, so the diameter of an opening matches the diameter of its maker. Green Woodpecker hole entrances are 6-7cm (2.4–2.8in) in diameter, Great Spotted 5-6cm (2–2.4in), and Lesser Spotted 3-3.5cm (1.2–1.4in). All three make round entrances, although those of the Green Woodpecker are sometimes slightly oval.

Great Spotted Woodpecker hole   Lesser Spotted Woodpecker hole   Green Woodpecker hole

Woodpeckers and honeyguides

A brood parasite is an animal that uses another to raise its young. Around the world various birds, fish and insects use this ingenious strategy, where the parasitic parents let someone else do all the hard work of building a nest and feeding the young. One of the most well-known brood parasites, and the only one in Europe to exploit other birds in this way, is the Cuckoo, or Common Cuckoo (*Cuculus canorus*). Cuckoos don't

parasitise woodpeckers, but there is a family of birds that do – the honeyguides; relatives of woodpeckers, placed by taxonomists in the same order Piciformes. They are famous as the birds that lead humans (and allegedly Honey Badgers) to beehives and wasp nests – hence their name. There are around a dozen species of honeyguide in Africa, two in Asia, and several use woodpeckers as foster parents for their young. The most widespread in Africa are the Lesser Honeyguide (*Indicator minor*) and Greater Honeyguide (*Indicator indicator*) and both

Above: Greater Honeyguides are regular brood parasites of woodpeckers in Africa.

are brood parasites of hole-nesters like woodpeckers. They do this by using egg mimicry, which is laying eggs that resemble the host's. They place one egg in one woodpecker hole, laying about five eggs in series (all in different nests). Then comes the grim bit. When hatched, young honeyguides puncture the hosts' eggs using a sharp hook on their beak. If they find themselves amongst any woodpecker nestlings, they either stab them or push them up and out of the hole. It is sometimes hard not to impose human attitudes on wildlife and start to harshly judge birds like honeyguides, but brood parasitism is just one of many strategies that has evolved to allow animals to pass on their genes to the next generation.

Eggs and incubation

A few days after mating, the female is ready to lay her eggs. She lays one egg per day until the clutch is complete. Woodpecker eggs are elliptical (oval) and plain white, with no camouflaging blotches or spots at all. This is typical for cavity-nesting birds, whose eggs lie at the bottom of a dark chamber away from the eyes of predators.

Green Woodpeckers usually lay five to seven eggs which are 31 x 23mm (1.2 x 0.9in). The clutches are incubated for 19 or 20 days.

Great Spotted Woodpeckers usually lay four to six eggs which are 27 x 20mm (1.1 x 0.8in). These are incubated for 14 to 16 days.

Lesser Spotted Woodpeckers usually lay four to six eggs which are 19 x 15mm (0.7 x 0.6 in). These are incubated for 14 to 16 days.

Left: In common with most birds that lay their eggs in dark cavities, woodpecker eggs are white. This is a clutch of Wryneck eggs inside a nest box photographed by researchers in Switzerland.

Wrynecks in mainland Europe regularly lay six to ten eggs; these are 20 x 15mm (0.8 x 0.6in) and are incubated for 12 to 14 days.

All four species lay their eggs in spring, which can mean any time from April to mid-June. The precise time varies from year to year depending upon several factors, but average temperatures in the preceding months are significant – very cold winters result in a delayed spring and laying dates.

Life in the chamber

Although woodpeckers incubate their eggs for a comparatively short time compared to most other hole-nesting birds, their offspring stay in the nesting chamber for a relatively long time. So much time and hard work is expended in hacking out the nesting cavity and bringing up the family that there is simply not enough time to produce another. They seldom start all over again if a clutch or brood fails or is lost, probably for the same reason.

Even the most fervent picid aficionado could not seriously describe woodpecker hatchlings as cute. Their eyes and ears are sealed, they are naked, with no proper down, and are unable to regulate their own body temperature. They are helpless, totally dependent on their parents – in ornithological terms, altricial. It's a simple case of function before beauty, but after the first few feeds, they set off on a course of rapid growth. They develop bulging white flanges at the corners of their mouths which help their parents locate their mouths in the dark nesting chamber. These flanges are sensitive, so when touched, the chicks instantly open their bills wide. Their limbs and claws look monstrous, oversized for their scrawny bodies, but this is because they will soon be needed for climbing the wooden chamber wall. Then there is the bill, which also looks too big, but which is going to be an important tool in their lives and the sooner it develops and they learn how to use it, the better.

Below: Deep inside the nesting chamber, a Wryneck feeds its young.

Parents working together

Woodpeckers are good parents. Both sexes incubate the eggs, brood (both have well-developed brood patches – featherless skin on the underside of birds during the nesting season) and feed the nestlings and keep the nest chamber hygienic by removing excrement. Nestling droppings come in convenient gelatinous sacs – disposable woodpecker nappies – which are carried away in the bill by the parents and then released at some distance. Although both parents carry out these duties, they do not do any of them together; they are done in parallel, as already mentioned. The males of the three resident British species are all involved in the breeding process from beginning to end; in fact, they usually do more of each job than females. Daytime chores may be shared equally, but males do the incubating and brooding overnight shifts; they are confident that they sired the offspring they are looking after, so they are keen to ensure they survive.

Above: A Wryneck feeds one of its chicks at the nest hole entrance. In common with other woodpeckers, both Wryneck parents cooperate in raising their young.

Left: A pair of Green Woodpeckers (female on left, male on right) attend one of their well-grown chicks who seems to be ready to fledge.

Time to get out and about

It must be quite comfortable in a woodpecker cavity because the young often seem reluctant to leave it; after all, they have had it easy, with their parents caring for their needs like chambermaids. As they put on weight and their bills get unwieldy, the cramped conditions force the issue and they fledge. A dose of curiosity is probably involved, too, as they begin to try out their climbing gear, scrambling up to the entrance hole and peering out. For a while they are fed at the entrance,

Below: Encouraged by its father, a young Great Spotted Woodpecker leaves the safety of the nest hole for the very first time.

before their parents start to ease up, visiting less often, but calling and sometimes gently drumming nearby to entice their young to come out for meals. Green and Great Spotted Woodpeckers fledge on average at about 22 days; Lesser Spotted Woodpeckers leave a little sooner at about 20. Wrynecks take the plunge at 18 to 24 days. Once out of the nest hole, fledglings follow their parents about (still demanding to be fed) for about three weeks in the case of Great Spotted Woodpeckers. It is known from radio-tracking that broods are split between the male and the female soon after fledging – a great insurance policy against total loss of the family. The first few days out of the nest hole are probably the most dangerous of a woodpecker's life – in suburban Britain cats are a serious threat now, as they are to all naive fledglings. Most parents feed their pushy offspring for a few weeks, sometimes longer, but by autumn the young are on their own and no doubt find that it is a harsh world out in the open.

Nest boxes

Most woodpeckers won't nest in manufactured nest boxes – remember they have an inborn need to excavate their own holes and are very choosy about where they nest. The Wryneck, which is unable to make holes, readily occupies hole-fronted nest boxes. The three resident British species will occasionally breed in nest boxes but are more likely to use them as roosts. Some Great Spotted Woodpeckers have learned that nest boxes are not only convenient to sleep in but also a good place to get an easy meal (see page 100). Nevertheless, if you wish to attract a woodpecker to nest near you, the Great Spotted Woodpecker is probably the best bet. Here are a few tips and specifications.

Left: As they are unable to excavate holes themselves, Wrynecks will readily use nest boxes to raise their young.

Below: Great Spotted Woodpeckers will occasionally breed and roost in nest boxes, but they also visit them to prey on any songbird chicks within.

- **Timing** Put the box up in the autumn or winter when woodpeckers are prospecting sites.

- **Type/sizes** Use a hole-fronted wooden box. Make one yourself (see the BTO website) or buy a ready-made 'Starling' box. Height 40–50cm (15.7–19.7in), width 16cm (6.3in), depth 18–20cm (7.1–7.9in), entrance hole diameter 4–5cm (1.6–2in).

- **Location** Place in a secluded spot, with several trees around, but with a clear flight path to the hole. Fix on a wide trunk with strong nylon tags or wire as nails can damage the tree.

- **Orientation** Tilt the front of the box slightly downwards and use one with a sloping roof to keep rain from the entrance hole. Face away from the prevailing wind direction.

- **Height** 3m–5m (9.8–16.4 feet) above the ground.

- **Filling** Fill the box with a block of soft wood, such as balsa, so the birds can hollow it out themselves. Leave the entrance hole partly excavated.

Communication

Although they are extraordinary birds in many ways, even their most avid admirers will admit that woodpeckers are not the best songsters. They make a range of vocalisations, but when it comes to really getting their messages across, most woodpeckers do it in another, remarkable way – a mechanical means of communication we call drumming.

Calls and songs

Woodpeckers don't really sing, but they do call. The difference between bird songs and calls is in their structures and functions. Calls tend to be short and simple and are mainly used as alarms, scolds and to keep in contact with mates and young. Songs are more complex, usually multinoted, long, with repeated phrases and used in courtship or to declare and defend a territory. The familiar songbirds of the UK – Robin, Blackbird, Song Thrush, Chaffinch, Blackcap, Dunnock – all sing for these reasons. Compared to birds like these, woodpeckers have a rather basic vocal repertoire, often just a few sharp high-pitched notes and some non-musical chattering, rattling and crying. Their most pleasant sounds are arguably the whistling that a few species manage and perhaps the Green Woodpecker's 'laughing'.

Above: With its head back and bill wide open, a young Green Woodpecker calls for its parents.

Opposite: This incredible photograph illustrates the speed with which woodpeckers hit their heads on trees when drumming.

Male and female woodpeckers generally use similar harsh calls in tense situations such as when they are threatened by rivals or predators. Mated pairs often indulge in a quiet *tête-à-tête* of gentle low-pitched notes and nestlings give away their presence by make continuous hissing, rasping or chugging begging calls from deep within the cavity. Interestingly, studies have shown that a sound frequency of between 1 and 2.5 kHz is characteristic for woodpecker calls, which is considered a perfect range for transmission in leafy woodlands. Now let's look at how the four British woodpeckers chat to one another.

Yaffling all the way

Green Woodpeckers are often heard before they are seen. They call somewhat randomly, but their 'laughing song' is one of the most familiar sounds of the British countryside – even non-birders seem to recognise it. Sometimes it's musical, sometimes chuckling, sometimes mocking, but always unmistakable. Males typically 'yaffle' in a series of up to 20 quite loud and clear *klu* or *plu* notes which speed up but fall away in volume, while females utter a thinner *pu-pu-pu*. In other situations, both sexes make sharp, single or double *kewk* or *kuk* sounds, a crying sequence of *kjack* notes and quiet clucking *gluk* sounds by the nest hole. When startled on the ground, they often take flight belting out a yelping, rapid *kjuck-kjuck-kjuck-kjuck!* stressed on alternate syllables. Newly fledged young are very vocal throughout July and into August making a double but frequently repeated *teu-tuo* call (inflected on the second note) as they maintain contact with their parents. This is even more apparent at a time of year when most songbirds are quiet.

Chick and chips

Below: Young Great Spotted Woodpeckers will beg incessantly for food from their devoted parents both when in the nest hole and when outside.

Great Spotted Woodpeckers are often noisy, and when excited or in a quarrel over food or a hole, they can get quite rowdy. Their most common call is a sharp, high-pitched *kick* or *chick*, often made twice but sometimes slowly repeated – in subdued moments it is softer, more like *chip*. When they are upset, these same notes are strung together in a rapid loud series or they make a harsh *krraarraarrr* alarm. They also chatter excitedly with many *chrett* notes, sometimes finishing off with a harsh rattle, particularly in early spring as birds vie for territory. When their parents are away, nestlings make a racket, uttering ceaseless thin *zizzing* sounds from the depths of the nesting chamber, but instinctively go quiet when they hear something unusual like scratching on the tree, presumably as they suspect a predator is around.

Little but loud

The most common call of Lesser Spotted Woodpeckers is a high-pitched series of 8–16 rapid, piping, shrill *kee* or *pee* notes that slow down towards the end. It's impressively loud for such a small bird and once learned is unmistakable. A variant of this is a slower sequence of a dozen or more *piit, piit, pitt* notes. When they feel threatened near the nest, both parents make brief, shrill *kick* or *pick* alarm call notes. Pairs and young keep in touch with each other with clucking *chuck,* clicking *chick* and squeaky *gig* notes.

Right: Despite being Britain's smallest woodpecker, male Lesser Spotted Woodpeckers can make quite a noise when they are calling for a mate.

Get lucky

Male Wrynecks are very vocal in spring when they arrive back in northern Europe from their winter holidays in Africa or the Mediterranean and are keen to find a mate – though these days their 'song' is not often heard in Britain. They call constantly from prominent perches with 10–20 loud, whining *tu* or *quee* notes, the whole phrase rising in pitch – not unlike a small falcon. So, a word of warning: if you think you have got lucky and are hearing a Wryneck, double check that it's not a Kestrel! Mated pairs make intimate cooing sounds near their nest hole, and when alarmed harsher *tek* or *tyuk* notes which vary in pitch. Nestlings make *zizzing* and *hissing* noises when disturbed or handled, for example by ringers.

Above: In spring, Wrynecks 'sing' from prominent perches. Sadly, this is a rare event in the UK today.

Drumming

Woodpeckers might not be great singers, but they are exceptional percussionists. Drumming is a non-vocal method of communication they use that functions very much as song does. It's how they communicate amongst themselves, warning rivals, showing off to potential partners, using it variously as a threat and also as an invitation. Let's make that clear: when woodpeckers are drumming, they are not feeding nor making nest holes and drumming does not damage trees nor buildings – a common misconception. They are simply pounding out a message to rivals that 'This is my place, can you hear how strong I am, stay away, or else', or to potential mates 'Hello! I'm here, why don't you visit and see what I can do?' As with calls, drumming sounds are low-frequency, ideal for long-distance broadcasts through woods and forests. High frequencies are often deflected by vegetation and generally don't carry far, whereas low-frequency sounds have bigger sound waves, capable of bending around trees, and so messages can be transmitted over greater distances.

As it is generally to do with territorial behaviour, woodpeckers drum most in spring before they start breeding. Increasing day length triggers both males and females to frequently drum, but just as male songbirds generally sing more than females, males tend to drum more often than females, and louder and more powerfully. Truth is, drumming is a rather macho thing. When setting up a territory at the end of winter, some males will drum hundreds of times a day, only easing off when they have a mate and all rivals have been banished.

Below: This Great Spotted Woodpecker is drumming on a dead, dry snag – an ideal substrate for transmitting messages over long distances.

Unpaired males carry on their sessions well into the spring. Apart from territorial issues a few soft rolls are sometimes done to lure any reluctant young out of the nest hole when the time comes to fledge, and some birds drum on the walls inside the nesting chamber, perhaps to tell their partner to come and take over duties.

The sounds produced by drumming are rather basic and uncomplicated when compared to most bird song, but birds do not just bang away haphazardly; drumming is structured. It is species-specific, with every species drumming in a different way in terms of the speed and length of rolls, number of beats and the time between beats. Studies of American woodpeckers have revealed that individuals produce their own pattern of vocalisation which scientists call 'vocal individuality' and it is reasonable to expect that our woodpeckers are no different. This probably means birds of the same species can identify each other and guess the condition of the drummer from the quality and intensity of their efforts, and potential partners and wily opponents can decide what moves to make.

Above: Wrynecks do not drum, although they sometimes lightly tap with their bill by the nest hole. This behaviour is known as 'demonstrative tapping' and, like drumming, is a way of communicating.

Not all species drum and of those that do, not all do so to the same extent. Some are fervent drummers, others very half-hearted. In fact, the drumming repertoire of many species is unknown and in some cases, it is not known whether they drum or not. Most woodpeckers, even those that do not drum, will tap lightly with their bills around the nest hole, particularly on the rim of the entrance and inside on the cavity walls. They seem to do this to show each other the nest, to encourage changeovers and perhaps simply just to bond with each other. This demonstrative tapping is not rhythmic like drumming but is another way they communicate non-vocally.

Drumming posts

Above: Some woodpeckers have worked out that drumming on metal objects, rather than on wooden surfaces, can be an even better way to get out their message.

Woodpeckers don't use any old tree to drum on. Within its territory, a male will have a favourite drumming post, a snag, dead limb or hollow branch, that resonates well when struck. Just as songbirds like to sing from prominent perches, drumming posts are often high up, in the open, away from the muffling effects of foliage. The whole point is to get the message out, loud, clear and far. To this end, urban-living woodpeckers often work out that using an artificial surface like a metal gutter, drainage pipe, weather vane or TV dish is better than using a wooden one – doing this certainly shows some brainpower (see page 50).

Waveforms of woodpecker drummings

Waveforms (oscillograms) are graphs that show the patterns – length, speed, rhythm (or cadence) – of woodpecker drum rolls. By looking at waveforms we can compare the differences between woodpecker species and potentially individuals.

Great Spotted Woodpeckers' amplitude peaks at the third strike, then falls away after midway. Strike intervals decrease towards the end. The longest interval is at the start (80ms) and the shortest (42ms) at the end. The roll lasts just under 1 second in this waveform.

The Lesser Spotted Woodpeckers' rhythm is very even, with strike intervals similar across the whole roll (47ms). Amplitude slowly rises and then diminishes – and so is often likened to a short machine gun salvo. The total length of the roll in this example is just under 1 second.

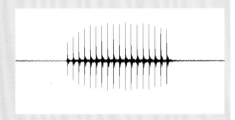

Green Woodpeckers do not drum loudly. The rhythm is even but less so than Lesser Spotted's. Amplitude varies, but is often hard to discern. Strike intervals are 47–50 milliseconds over the first half of the roll, slowing to 77 milliseconds at the end. The roll shown here lasts about 1.1 seconds.

The British drummers

No, we are not ranking Ringo Starr, Keith Moon or Ginger Baker here (apologies to the younger generation reading this) but Great Spotted, Lesser Spotted and Green Woodpeckers. Wrynecks don't drum at all, they just don't have the biological kit for it. Green Woodpeckers don't bang the drum very often, and when they do, it is usually the male who knocks out some soft, feeble rolls of 1–2 seconds long, usually close to the nesting site.

Sometimes a few unconvincing rumbling bouts are done between calling. Great Spotted Woodpeckers, on the other hand, are impressive percussionists. Typical Great Spotted drums are short (less than one second long) but explosive bursts of 10–16 strikes that speed up before the end. Males often start to perform early in the year, on crisp, clear winter days, and by the time spring arrives, eager individuals may drum hundreds of times per day. Between 6 and 9 sessions per minute have been recorded, but this stops once the eggs are laid and the chores of breeding take over. Females also drum, perhaps to solicit males or warn off other females, but their efforts are shorter and more subdued. Lesser Spotted Woodpeckers are first-rate little drum majors. Both sexes drum, but, as is so often the case, the males tend to perform more often and better. Rolls are speedy, high-pitched and drawn out, but with an even tempo, often 20–30 strikes over 1–2 seconds. It's not unusual for a second roll to quickly follow a first and to be interspersed with calling. Rolls often sound rattling because they are performed on thin snags or branches.

Below: Male Lesser Spotted Woodpecker. In spring, Britain's smallest woodpecker is often heard drumming and calling before it is seen.

International drummers

Around the world, a rough rule is that the ground-loving species, like our own Green Woodpecker and many in the *Campethera* and *Dendropicos* genera in Africa, drum less and relatively poorly compared to their tree haunting relatives. You'd be forgiven for thinking that the bigger the woodpecker, the better the drummer, but that is not the case. The large *Campephilus* woodpeckers of Central and South Americas make simple double or triple raps or knocks rather than proper rolls, and the biggest woodpecker on the planet, the Great Slaty Woodpecker, doesn't beat about the bush at all, and never drums. Some species have developed very specific styles of their own: in Asia, the Rufous Woodpecker makes distinct knocking rolls that slow down and grind to a halt like a stalling old motorcycle. In North America, the sapsuckers all knock out irregular rolls of uncertain rhythm that decrease in beats as they progress and end with single or double taps.

Above: The gorgeous Crimson-bellied Woodpecker is a woodpecker that communicates with knocks rather than drum rolls.

Left: Both male and female Rufous Woodpeckers make diagnostic drum rolls of up to five seconds long that sound like a misfiring motorbike.

Other drummers

Woodpeckers are not the only drummers of the bird world. The aerial display of a male Common Snipe (*Gallinago gallinago*), where he rises and then dives, letting the air whistle through and vibrate his fanned out outer tail feathers, is also called drumming (as well as winnowing and bleating). Then there is the Palm Cockatoo (*Probosciger aterrimus*) which lives in the rainforests of Indonesia, New Guinea and north-east Australia. This ingenious parrot is the only animal on the planet, apart from humans, known to make and use a tool to make music. Males hold a seed pod or twig like a drumstick and bang it rhythmically on a tree hollow in front

of a female while calling and hopping about. Woodpeckers may well be the best drummers in the bird world, but the Palm Cockatoo can sing and dance as well.

The Importance of Woodpeckers

Woodpeckers are not just members of the woodland and forest communities where they live; they are also animal architects that shape and influence their environment for themselves and for many of the animals they live alongside. As we shall see, woodpeckers also benefit people – in perhaps unexpected and surprising ways.

Keystones and indicators

Remove the keystone from the top of an arch and the whole structure will come tumbling down. Simply put, a keystone is something on which other things depend for support. Ecologists have borrowed the word 'keystone' from masons and use it to mean a species that plays a vital role in an ecosystem, by helping it maintain its structure and diversity. Keystones modify habitats in ways that affect the other animals that live there. There are many cases where species were only recognised as being keystones when they disappeared or were removed from an ecosystem and the resulting negative changes to the community left behind was noticed. Woodpeckers are keystones because they alter the wooded habitats around them by creating feeding opportunities for others, controlling insect populations and, very importantly, by making cavities that many other animals then use. It's likely that the 'new wave' of mammals, which thrived after the

Opposite: The Edible Dormouse is one of the many animals that make woodpecker holes their home.

Below: This experiment seems to prove that Green Woodpecker tongues can reach the grubs that most other bird tongues cannot reach!

Above: Ancient gnarled and stunted oak trees growing amongst mossy boulders in Wistman's Wood, a remote upland oakwood and National Nature Reserve in Dartmoor National Park, Devon. Very few wooded habitats like this remain in Britain.

dinosaurs, benefitted from the woodpeckers that exploited the vast primeval forests, providing them with shelter in the cavities they made, and it has been like that ever since. Woodpeckers are in effect ecosystem engineers. The fact that they unintentionally supply holes for others does not matter – take away the woodpecker and the woodland community starts to crumble.

Another term used by ecologists is 'indicator species'. This is an animal that demonstrates the quality of a habitat by simply being there. When salmon and otters are spotted in a river where they have not been seen for years, you know that river is healthy again. The same goes for woodpeckers and woodlands – you don't see many in monotonous plantations of non-native conifers or eucalyptus. It's quite simple: woodpecker diversity is a first-rate indicator of the overall biodiversity and health of woods and forests because if they are around and doing well, the habitat is probably in good condition, too.

Primary and secondary

Woodpeckers are 'primary cavity excavators', birds that make their own nesting and roosting holes. Those animals that use holes for nesting, denning or roosting, but cannot make them, are called 'secondary cavity users'. As primary cavity excavators, woodpeckers usually create most of the tree holes found in an area but then abandon them after their first use. The secondary cavity users then move in. In Britain Nuthatches, Starlings, tits, flycatchers and even bees and wasps all use old nest holes to breed or roost. Studies on woodland bats in England have shown how important woodpeckers are in the 'cavity web' with species such as the red-listed Bechstein's Bat (*Myotis bechsteinii*) being dependent on their hole-making activities for their maternal roosts. Elsewhere in the world, we can add ducks, hornbills, owls, parrots, pigeons, rollers, toucans, trogons and many others to the list of birds that benefit from woodpeckers.

Above: Starlings are just one of the birds that nest in holes in trees but which are unable to make them, so they regularly use those made by woodpeckers.

Below: A young Grey Squirrel making use of a woodpecker hole in north Yorkshire, England.

In plantations and overly tidied woodlands, there is often a dearth of natural holes (created by weather and decay) in trees and this can seriously limit nesting possibilities for secondary cavity users. It is not just a case of using a woodpecker hole because there is little else around; their holes are high quality properties, designed as the best homes in which to raise a family. They are in great demand; indeed, some birds that can make holes, like Nuthatches and tits, still prefer to use them. Today the provision of nest boxes often helps solve the hole availability problem, but this does not detract from the keystone role of woodpeckers.

Natural pest controllers

Right: Studies in North America have shown that Downy Woodpeckers (which are about the same size as our Lesser Spotted) are important natural controllers of non-native insect pests that can otherwise devastate forests.

Below: Across Europe the Three-toed Woodpecker is a friend of foresters as it mainly preys upon Spruce Bark beetles an insect pest that munches and ruins timber.

An increasing number of studies from around the world are proving just how important woodpeckers are at keeping forest insect pests in check. As most woodpeckers don't migrate, they also do this all year round. Modern pesticides are just that, modern, meaning new – woodpeckers have been preying on the very beetles that foresters loathe for eons. In North America, Black-backed (*Picoides arcticus*), Red-bellied (*Melanerpes carolinus*), Hairy (*Picoides villosus*), the tiny Downy Woodpecker and others are being increasingly recognised as natural biological controllers of wood-boring beetles such as the Emerald Ash Borer, an alien invader that has gnawed its way through millions of ash trees across the Great Lakes region. In Europe, Three-toed Woodpeckers prey mainly on Spruce Bark Beetles, an insect which can devastate commercial conifer forests. In India, the Rufous Woodpecker eats the ants which foster the mealy bugs that are harmful to coffee plants. Since this was made clear, an increasing number of enlightened growers are happy to see as many of these woodpeckers in their plantations as possible. Why spray with toxic and expensive chemicals when the woodpeckers will do the job sustainably and for free?

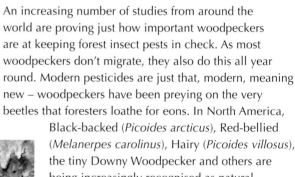

Biomimicry

Inventors and designers are increasingly looking to the natural world for ideas and sustainable solutions when designing materials, structures and systems. This approach of copying nature's strategies and patterns is called biomimicry. The shock-absorbing anatomy of woodpeckers (see page 33) has been studied to see how improvements can be made to all kinds of devices and gadgets such as the 'black box' that houses aircraft flight recorders. Research has shown that the head of a hammering or drumming woodpecker can withstand accelerations some 16 times greater than humans can without suffering concussion. Doctors studying human dementia and Alzheimer's disease have commented that people could never be exposed to the kind of constant pounding of their heads that woodpeckers do and that it is a wonder that woodlands are not littered with dazed and dying woodpeckers. Hence, woodpecker heads are now being examined by neurosurgeons investigating brain injuries such as CTE (chronic traumatic encephalopathy) which are sustained by athletes in contact sports. It is thought that the designs of the safety helmets used by American football players and others could be improved by copying woodpeckers. Eye surgeons are also looking at how the designs inside woodpecker skulls might be used in the treatment of detached retinas.

Left: The remarkable shock-absorbing properties of woodpecker skulls and heads have been studied, copied and incorporated into, amongst other things, the helmets worn by American footballers.

Threats

Above: The sad sight of a Sunda Woodpecker caged in a market in Java. Indonesia's trade in wild birds, some legally trapped but most illegally, is devastating the country's natural heritage.

Below: An illegal logging camp in Sumatra, Indonesia. The ongoing, rampant felling of the world's tropical forests is not only a tragedy for woodpeckers, but also for all wildlife.

The deliberate killing of woodpeckers by people is unusual as they don't raid chicken coops, are not grouse-eating 'villains' and apparently don't taste that good themselves, so they tend to escape the gun and pot. In any case, it would be illegal to kill any woodpecker in Britain as they, like all wild birds and their eggs and nests, are protected under the law. Woodpeckers are killed, like many animals, in indirect ways. They suffer, and ultimately disappear, as the places they live in are ruined by modern farming methods such as the overuse of herbicides, pesticides and fertilisers, the removal of hedges and copses and intensive forestry practices. The management of forests is inextricably linked to forest bird diversity; the more intensive, the worse it is. For example, clearfell harvesting, which removes all standing trees before they reach an age when they become useable to woodpeckers, clearly limits the availability of potential nest sites. Fortunately, there is growing interest across Europe in 'close to nature' forest management, which allows trees to develop to maturity. Many forest managers today recognise the benefits of a better understanding of the forest ecosystem, which includes woodpeckers. Survival is the name of the game for all wildlife, and it's often tough. Sadly, when commercial interests conflict with those of wildlife, wildlife generally loses.

Which woodlands?

Britain's woodpeckers need trees, right? Of course, but they need *certain kinds* of trees. Although forest cover in Britain has increased over the last hundred years, most of the new woodlands are plantations of non-native conifers – which support fewer insects than native woodlands and hence fewer insect-eating birds. The loss of Britain's ancient native woodland, the fragmentation and isolation of that which remains, and the disappearance of orchards (a staggering 90 per cent have gone since World War Two) have been key issues in the decline of Lesser Spotted Woodpeckers, but things are far from clear. We seem to know what Lesser Spotted Woodpeckers like, but we don't fully understand why they have declined to the extent they have. Woodpeckers love standing, dead or dying trees and bare upright branches (snags). Strange as it may seem, dead timber is often better for wildlife than living trees, possibly as it holds more wood-boring insects; therefore, the more snags there are in a woodland, the more woodpeckers and other cavity-nesters benefit. In Britain, unfriendly woodland

management and 'health and safety' concerns often meant snags were removed without enough thought as to their value. Many woodlands are just too tidy, which is bad news for woodpeckers. Then again, there is good news, too. Since the 1980s the Woodland Grant Scheme has offered incentives to landowners and foresters to plant indigenous trees – particularly broadleaf rather than conifer in lowlands – and slowly but surely the amount of better tree cover is increasing. If this is coupled with a more determined move towards 'close-to-nature' management, the future for our woods and woodpeckers will be brighter. Native broadleaved woodlands are best, even for the adaptable Great Spotted Woodpecker.

Above: Britain's native woodlands are places of beauty, but sadly, we have degraded or felled many, and birds such as the Lesser Spotted Woodpecker have consequently suffered.

Alien enemies

Above: Found only on the Japanese island after which it is named, the Okinawa Woodpecker is one of the rarest woodpeckers on Earth.

Below: It has been estimated that pet cats kill at least 55 million birds a year in the UK. Green Woodpeckers, like this one, being ground feeders, are probably easy prey for any feline.

Woodpeckers are, of course, sometimes taken by predators, such as the Sparrowhawk (*Accipiter nisus*), but the four British species are not the main prey of this raptor, nor of any other hunter. Nesting and roosting in tree holes, out of reach of most predators, certainly helps. An enormous threat to wildlife worldwide is that posed by non-native species, introduced by people, unintentionally or deliberately, into places where they do not naturally occur. Wildlife on islands is particularly vulnerable as it is often unafraid of the new arrivals or has ineffective defences against them. The Okinawa Woodpecker (*Dendrocopos noguchii*) lives only on the Japanese island from which it gets its name. With perhaps under 100 breeding pairs, it is listed as *Critically Endangered*, having suffered from numerous threats including predation by introduced mongooses. Closer to home, there is the problem of the millions of pet cats roaming around suburban Britain. Domestic cats are instinctive hunters and it is known that they kill garden birds in large numbers. The most common victims are ground-feeding songbirds like Blackbirds and Robins, but woodpeckers are at risk, too, when they venture into gardens to take suet and seed. Cats killing birds in Britain is, sadly, a daily event, but one incident from a few years ago made the national press: a Wryneck turned up in Burnham-on-Sea in Somerset (where it is an extremely rare visitor) only to be promptly pounced on by a pet cat.

Conservation

The International Union for Conservation of Nature (IUCN) focuses on nature conservation and the sustainable use of natural resources. It maintains a 'Red List' that classifies species depending on their global extinction risk. The highest levels of threat are *Critically Endangered* (CR), *Endangered* (EN), and *Vulnerable* (VU). None of Britain's woodpeckers are threatened by extinction – fortunately there are still plenty of Lesser Spotted Woodpeckers on the European continent. The UK has a 'Red List' of its own with three status categories – Red, Amber, Green. Green and Great Spotted Woodpeckers are both Green, but Lesser Spotted is Red. The Wryneck is now such a rare breeding bird it does not even get categorised.

Above: Wrynecks were once not uncommon in Britain, as many folk names and superstitions involving the bird testify to, but today they are very rare breeders.

Globally, 18 woodpecker species are included in one of the above IUCN categories. Most woodpeckers are arboreal and it's clear their decline is directly linked to people's insatiable need for timber. The worst affected areas are tropical rainforests, in countries such as Brazil, Haiti, Indonesia, Malaysia, Nigeria and the Philippines. In some of these countries over 50 per cent of original forest cover has gone – dreadful figures by any standards.

Above: It is believed that if the current rate of rainforest logging continues, in 100 years there will be none left.

Closer to home things are not too rosy either. In 2016 the RSPB and other nature conservation bodies published a report entitled *State of Nature 2016*. This comprehensive report presented data on over 3,000 species of animal and plant found in Britain. The conclusion was not easy reading – around 60 per cent of species had declined in the study period. On the woodpecker front, the Lesser Spotted Woodpecker was found to be in the most trouble.

Above: The British conservation status category of Lesser Spotted Woodpecker is now Red.

Damage limitation

Above: Woodpeckers do not endear themselves to telecom companies when they drill utility poles with holes. Even creosote does not always deter a determined woodpecker.

Below: Holes made by Great Spotted Woodpeckers in the Tower Hide at Wicken Fen, Cambridgeshire. Note the attempts to patch up the holes!

Woodpeckers are wonderful birds; everyone seems to like them. They certainly have character and their carpentry skills are inspirational, but they can upset some people and are sometimes even called pests. Green Woodpeckers sometimes break into beehives and woodpeckers 'attacking' historical wooden buildings in England occasionally make the papers. Both Green and Great Spotted Woodpeckers probably find old (and no doubt rotting) cedar shingle roofing easy to peck. One of the most infamous records was when Great Spotted Woodpeckers made around 200 holes in the wooden bell tower of a Grade One listed 1,000-year-old church in Essex.

They also hack holes in wooden electricity poles. An estimated 6,000 were damaged in south-east England in 2017 and, as Great Spotted Woodpeckers spread northwards, this is being increasingly reported in Scotland – often by annoyed telecom and energy companies who foot the repair bills. Quite why woodpeckers excavate man-made wooden structures is not entirely clear. Certainly, old churches infested with insect colonies will be attractive, but why creosote-treated poles seemingly devoid of prey are drilled is something of a mystery. In some places, a lack of suitable trees and snags in which to make roost holes may be to blame.

Great Spotted Woodpeckers are also sometimes seen as villains because they plunder the nests of other hole-nesting birds and eat the helpless chicks. Grisly to see perhaps, especially when it happens at a nest box you have put up for songbirds in your garden, but studies have shown that this behaviour does not have any major effect on the overall populations of the victims.

Troublesome woodpecker?

When you have a 'problem' woodpecker, the first thing to do is understand what it is doing. Is it making a nest or roost hole or looking for food? If it is nesting, there is not much you can do because in Britain all woodpeckers are legally protected, and it is illegal to harm them or disturb their nests, even while woodpeckers are still building them. Prevention is the answer. Here are some ways woodpeckers can be gently discouraged from causing damage.

What to do

- Erect visual scares like wind-socks, balloons, scare-eyes, flash tape, foil strips and reflective discs.
- Coat wood in a compound of aluminium ammonium sulphate to discourage woodpeckers from pecking it – it's available from most garden centres, but read the manufacturer's instructions for suitability of use.
- Metal plates around nest box entrance holes sometimes stop woodpeckers widening songbird-sized holes and getting at the eggs or chicks inside – though crafty woodpeckers often simply break in elsewhere. Alternatively, woodcrete nest boxes can prevent woodpeckers from accessing nest boxes occupied by other birds.

What not to do!

- Never try to catch an offending woodpecker.
- Never apply sticky or greasy products to the wood the birds are pecking – these can coat a woodpecker's plumage.
- Never use 'anti-woodpecker' paint – widely available on the internet – unless you are absolutely sure it is non-toxic.
- Don't bother with plastic owls or hawks as woodpeckers soon work out what they are and ignore them, or even peck them!

Finally, we might do well to remember that any 'trouble' or 'damage' woodpeckers cause is insignificant compared to the harm inflicted upon them and the places where they live.

Woodpeckers in Culture

Woodpeckers appear in mythology and folklore everywhere they occur. They have been respected and even revered by indigenous tribes in the Americas, Africa and Asia, all of whom have their own woodpecker legends. In Europe, too, they have captivated people, featuring in superstition and tales, and in ancient Greece and Rome they were almost deified.

Woodpeckers have been idealised and occasionally idolised as nature's master carpenters, tree surgeons and as guardians of wooded places. Nothing too unexpected there perhaps, given their lifestyle, but they have also been looked upon as talismans in war, weather forecasters and credited with teaching mankind how to make dug-out canoes and play the drums. They have been associated with great strength, bravery, fertility, prophecy, magic, medicine, friendship and happiness. The fact that woodpeckers make safe and sheltered tree holes for their family, and often successfully raise all their young, has seen them imbued with human qualities, admired as good home makers and dependable parents. In folk tales and legends worldwide, they are cast as heroes: being faithful, wise, wily, generous and modest. Then again, on the negative side of things, they also appear as scoundrels, and are variously fickle, foolish, devious, miserly and vain. Symbolically multitalented, woodpeckers have been just about everything people wanted them to be.

Dendrocopus major

Above: A male Great Spotted Woodpecker graces a stamp from the island of Jersey.

Opposite: A sculpture of a woodpecker on a tree trunk, Dumfries Galloway.

Native Americans

The vast damp forests of the Pacific Northwest of America are wonderful for woodpeckers – 13 species occur – and they feature in the spiritual world of the region's native peoples. Many Native American tribes valued all wildlife and continue to do so, believing in a relationship of reciprocity rather than exploitation.

Woodpeckers were medicine birds for some Native American tribes and their bills, tongues, feathers and red scalps were used on shamans' rattles, tomahawks and war bonnets and carried as amulets to ward off disease and evil spirits. Others used their scalps as currency – the Karuk calling the Pileated Woodpecker the 'dollar bird' because they could trade its red head feathers for one dollar. The Gitxsan and Tsimshian tribes carved huge-beaked woodpeckers into their totem poles.

Right: A giant legendary woodpecker of the Gitxsan in British Columbia in Canada, called Wee-Get-Welku, looks down from the top of a totem pole at an eagle.

Many Native American tribes held World Renewal ceremonies to ensure bountiful harvests and abundant game and to avert earthquakes, floods, disease and the end of the world – woodpeckers played a crucial part in these. The Hupa and Karuk ceremony was called the 'Woodpecker Hat Dance' where men wearing headdresses adorned with red scalps chanted and nodded their heads like woodpeckers pecking. Red was hugely important spiritually, being the colour of blood and therefore life and death. Cherokee warriors admired woodpeckers as brave birds and took their red heads into battle.

Further south, in the Sierra Madre mountains of Mexico, the Huichol people believe that a gigantic woodpecker once protected the sun, which humans had created, from the animals of the night that did not want it to shine. The woodpecker is finally killed by the jaguar and the wolf, but to this day its descendants have bright red crests, like the setting sun, to remind us of its bravery. The Mayans had a much simpler explanation for how the woodpecker got its red crest – their earth goddess *Chibirias* painted it on the bird with her brush.

Above: Hupa Indians in north-western California prized the red scalps of woodpeckers, which they associated with wealth and displayed at ceremonies for good fortune.

Drum and beat

The drumming of woodpeckers has frequently held a deep symbolic meaning for people. Woodpeckers were totemic birds, revered as the first drummers who taught mankind how to communicate with each other by using wooden drums. Their drumming was said to be in tune with the rhythm of the earth itself and shamans used their ceremonial drums to tap into this energy. The Bribri people of Costa Rica and Panama say the first drums were made by a woodpecker spirit, and in the Caribbean, Taíno tribes revered a sacred woodpecker called *Inriri Cahubabayael* that was said to have taught them how to beat on drums made from hollow logs. The rumbling sound of drumming has often been associated with thunder – the Minoans and then the ancient Greeks associated it with Zeus, the Father of the Gods, and his thunderbolts, and in Norse mythology drumming was linked to the thunderous hammer of Thor.

Touched by the woodpecker

In the past, the Ainu from Hokkaido in Japan fished and hunted using dug-out canoes. They called the woodpecker the '*boat-making bird*' as it was believed to have been brought to earth to teach people how to hollow out logs to make these canoes. On the other side of the world, in Central America, the Bribri people still use wooden dug-outs and say that anyone who is skilled at carving them has been 'touched by the woodpecker'.

Right: The Bribri people of Costa Rica and Panama, who make dug-out canoes like this one, admire the woodworking skills of woodpeckers.

There was a time when there were no bird books, no checklists, no websites, and very few taxonomists, so people made up bird names. These folk names were usually based on what the bird looked like, what it did, or the sounds a bird made (onomatopoeic). A good example of this is the Green Woodpecker whose mocking and laughing call inspired such wonderful English local names as *Yaffingale* (south-west), *Yelpingale* (Berkshire), *Yappingale* (Somerset), *Yuckle* (Wiltshire), *Yockel* (Shropshire), *Yaffler* (southern counties) *Yaffle* (widespread) and simply *Laughing Bird* (Shropshire).

Singing in the rain

Left: In several cultures, the Green Woodpecker was said to be a forecaster, or even a bringer, of rain.

Besides impressing the locals with its laughter, the Green Woodpecker was also commonly believed to be a weather forecaster. When it called, a storm or shower was surely on the way and it was dubbed the *Rainbird*, as well as *Rain-fowl*, *Rain-pie*, *Wet-bird*, *Weather-hatcher* and *Weather-cock*. These names varied from county to county and were used for other birds, too. This belief was not confined to Britain: all over Europe, from Scandinavia to the Mediterranean, woodpeckers (not only the Green) have been associated with the weather, especially bad weather, and most languages have their equivalents of 'rainbird'. The Romans called the Green Woodpecker the *pluviae aves* and it might be the case that they brought this notion north with them as they conquered Europe, although this is open to discussion, as both the Druids and ancient Welsh described the bird in a similar way.

Classical woodpeckers

Above: Three Great Spotted Woodpeckers can be found in this 17th-century oil painting of Romulus and Remus by Peter Paul Rubens.

Woodpeckers were not only rain forecasters for the Romans; they were said to be oracles, able to predict many of the things to come, and seeing or hearing one before a battle was a good omen. Politicians, generals and farmers all respected and protected them, in fact it was a crime to harm a woodpecker. The ancient historian Plutarch wrote '*Esteemed holy to the god Mars; the Latins still especially worship and honour the woodpecker.*' Above all, a woodpecker was believed to have helped the she-wolf in caring for the infants Romulus and Remus, the founders of Rome itself – and for a bird in mythology, there are very few roles better than that. Of course, the Romans absorbed many things from the cultures that came before them, including their woodpecker cults and myths, so let's look a little earlier in history.

Sometime between the ninth and third centuries BC, in what is now the Marche region in Italy, a pre-Roman people called the Piceni (also known as the Picenes) adopted the Green Woodpecker as their tribal totem after one was said to have helped them to find a home in the region. They became the 'people of the woodpecker' and lived in Picenum. We might wonder what they would say if they could only know that many of today's woodpecker terms – *Piciformes*, *Picidae*, *Picinae* and *Picus* – are derived from their tribal name, and that its image now appears on the coat of arms of Marche and the badge of a local football team.

In Greece, in Aristophanes' play *The Birds*, a woodpecker ruled the world until Zeus took its place. Ares, the God of War and Agriculture, was associated

Above: Even today, a modern, stylised woodpecker is at the centre of the emblem of Italy's Marche region.

with it and some classic scholars say that Pan, the half-man half-goat God of Nature, was hatched from a woodpecker's egg. Metamorphosis (change from one form to another) was a common form of punishment for anyone who offended the gods in ancient Greece: here's one story with a woodpecker ending. King Celeus of Eleusis and his men decided to enter a cave on Mount Ida to steal honey – not a good idea, as that cave was the sacred birthplace of Zeus who caught them in the act and as punishment changed them into birds, with Celeus becoming a Green Woodpecker. Roman myth, too, has its share of characters who were, or ended up as, woodpeckers, the most famed being King Picus who was transformed into a woodpecker by the sorceress Circe. His mistake was to reject Circe's advances after she declared her love for him, instead staying faithful to his wife.

Left: In ancient Greek mythology, woodpeckers and Zeus, the king of the gods, are inextricably linked.

The Barley Bird and Cuckoo's Mate

Above: According to the folklore of some rural areas in southern England, the Wryneck and the Cuckoo were thought to be friends because they arrived in the spring at roughly the same time.

As previously mentioned, Wrynecks used to be more common in Britain. Besides ornithological records, this can be seen by the many folk names, often local, that were used for the bird. *Writheneck* and *Snakebird* obviously refer to the Wryneck's habit of twisting its neck (see page 48), but in southern England it was called the *Barley Bird* as it arrived just as that cereal crop was being sown. *Cuckoo's Mate* – and variations on this theme such as *Knave, Footman, Messenger, Maid* – also refer to the bird's arrival in Britain, at roughly the same time as the Cuckoo. Sadly, times and conditions have changed and Wrynecks and Cuckoos rarely meet in Britain today.

Woody Woodpecker

Popular Western culture, Hollywood to be precise, has seen to it that even if you have never seen a real woodpecker, you've probably heard of one. In the 1950s, arguably the world's most recognisable *picid* appeared on TV screens in the USA and then across the globe in the *The Woody Woodpecker Show*. Today Woody even has his own star on the Hollywood Walk of Fame. Why create a woodpecker? Well, the story goes that his creator Walter Lantz was bothered by an Acorn Woodpecker that was loudly hacking holes in the cabin where he and his wife were on honeymoon. Lantz decided to get his gun, but Mrs Lantz persuaded him to spare the noisy bird and base his next cartoon character upon it. So, this wacky woodpecker, with his outlandish plumage, infuriating laugh, roguish behaviour and 'Guess who?' catchphrase, was hatched. The strange thing is, Woody's red crest and elongated body suggest a Pileated Woodpecker rather than an Acorn Woodpecker. Then again, his mostly blue feathering rules out all real woodpeckers as none on the planet has blue plumage. Woody does peck and drum, but most of his behaviour is very un-woodpecker-like. In 2017, Woody made a comeback in an American–Canadian film where he clashes with a lawyer who wants to build a house in a forest, only to find out he is cutting down a tree in which the intrepid woodpecker lives.

Above: The mischievous and at times downright wacky Woody Woodpecker.

Below: A Pileated Woodpecker – the real wild woodpecker species that perhaps most resembles the cartoon Woody.

Professor Yaffle

Though certainly less famous than his American cousin
Woody, another woodpecker made his mark on television
in the 1970s. Readers of a certain age might remember
with some nostalgia the BBC children's series *Bagpuss*.
Amongst its odd assortment of characters was a rickety,
wooden woodpecker bookend called Augustus Barclay
Yaffle. Complete with spectacles perched on his bill
and an aloof attitude, Professor Yaffle, as he was more
commonly known, was scholarly, with an explanation
for everything. Indeed, this clever old bird is said to have
been based upon the adroit English philosopher Bertrand
Russell. If *Bagpuss* was before your time, a quick trawl of
YouTube will reveal some episodes.

Weaselpecker

In March 2015, a woodpecker unexpectedly hit the headlines. A couple was walking in Hornchurch Country Park in Essex when they spotted a Green Woodpecker on the ground. Nothing unusual in that until they heard the bird squawk in distress and then suddenly take to the air with a weasel, which had attacked it, riding on its back. The quick-thinking gentleman managed to snap several photographs of the event which went viral on social media.

How and Where to Watch Woodpeckers

After reading about woodpeckers, you will now hopefully be keen to get out into the woods and hear them calling and drumming and, all being well, to see one going about its woodworking business.

There are very few places in Britain where there is a chance of seeing all four of the country's woodpeckers. Great Spotted Woodpeckers can be found just about anywhere with trees, and Green Woodpeckers are not too hard to find, especially in southern England, but Lesser Spotted Woodpeckers have become increasingly rare and Wrynecks are now just seasonal visitors. Generally, the south is the region to see all three residents, as well as having a chance of catching up with a migrating Wryneck. The New Forest hosts all three breeders, as does the RSPB's Minsmere reserve, and the Arne RSPB reserve in Purbeck, Dorset is another good bet all year round. Scotland is not woodpecker-rich, with only Great Spotted being widespread; Green Woodpeckers are thin on the ground, and Lesser Spotted Woodpeckers are absent. However, the best chance of seeing a Wryneck in Britain in the breeding season may well be in Scotland, in Deeside, Speyside or Strathspey, where pairs occasionally still nest. However, without inside knowledge and a few field skills, some detective work will be needed.

Opposite: A female Great Spotted Woodpecker posing on a dead branch at the RSPB's The Lodge nature reserve in Bedfordshire.

Below: Woodland at the RSPB's Minsmere nature reserve in Suffolk. This reserve is one of the few places in Britain where it's possible to see three woodpecker species in a day.

Great Spotted Woodpecker

By far the most common and widespread of Britain's woodpeckers, with an estimated 140,000 pairs, Great Spotted Woodpeckers are not hard to find. They are increasing their range, in fact, moving into northern Scotland where they were once uncommon. The reasons for this success are not entirely clear, but they are adaptable birds that even do well in suburban gardens – perhaps you have one visiting yours? The many people now feeding birds certainly help Great Spotted Woodpeckers. After a long period of absence, Great Spotted Woodpeckers have also recolonised Ireland, first nesting in Northern Ireland in 2006 and in the Republic of Ireland in 2009. Still rare, but steadily increasing, the Great Spotted is now established in the east. Regular sites in Northern Ireland include Belfast's Belvoir forest park, Tollymore forest park in Newcastle, Quoile Pondage reserve in Downpatrick, and Mountstewart House, Newtownards, all in County Down. In the south, the best areas to search are old woodlands at Glendalough in Wicklow, and in Louth, Meath, Wexford and Dublin.

Below: Great Spotted Woodpecker is the most common and widespread member of the family in the British Isles and indeed all across Europe.

Lesser Spotted Woodpecker

Britain's smallest woodpecker has probably never been common, but now it is in real trouble, officially in serious decline. Local losses were first noticed in the 1980s, but a nationwide collapse is now sadly obvious. The BTO estimate that numbers fell by 50 per cent in the decade before 1999. Today's breeding populations, perhaps just 1,500 pairs, hang on in England and Wales, and worryingly these are often isolated from one another. The reasons for this little woodpecker's troubles are unclear, but it is likely that the degradation and loss of Britain's woodlands are mainly to blame. It is now the case that a special trip is probably needed to watch this gem. Regular spots include Sizergh Castle in Lancashire, Moore NR in Cheshire, the Lugg Valley in Herefordshire, Santon Downham in Norfolk and the New Forest. In Wales, the RSPB's Ynys-hir reserve is perhaps the best bet.

Above and left: Old broadleaved woodland at the RSPB's Ynys-hir nature reserve in Ceredigion, Wales – home to both Great Spotted and Lesser Spotted Woodpeckers.

Above: Open gardens offer ideal conditions for Green Woodpeckers.

Green Woodpecker

Our largest woodpecker is widespread across much of southern England, but becomes patchier in the north, in west Wales and south-east Scotland. They are missing from much of western and northern Scotland, the Isle of Man and Ireland, and, as they are primarily sedentary, seldom roaming seasonally, there is not much chance of stumbling upon a nomad. The BTO estimates 52,000 pairs overall. Although mostly a bird of open woodlands, in recent decades Green Woodpeckers have spread into urban areas, visiting parks and gardens and farmland where lawns and grasslands are untreated and ants, their favourite food, are plentiful.

Wryneck

Wrynecks were last recorded breeding in England, in Buckinghamshire, in 1985. The odd pair has nested in Scotland since then, particularly in Strathspey, but to keep them safe from disturbance the exact sites are not publicised. Wrynecks probably disappeared as regular breeders in Britain due to changes in grassland farming practices – particularly the use of pesticides – and the loss of traditionally managed orchards, pastures and meadows. With declines also noted elsewhere in Europe, it is hard to see Wrynecks recolonising Britain anytime soon. It is estimated that around 200 individual Wrynecks drop into Britain each year when on migration southwards from Scandinavia, mostly from the end of August to October. They can turn up almost anywhere, but sites worth visiting are Portland in Dorset, Blakeney Point in Norfolk, Spurn Point in Yorkshire, Holy Island in Northumberland, Land's End, St Mary's Island in the Isles of Scilly and in Scotland the Isle of May.

Below: The Isle of May in the Firth of Forth in Scotland is a regular stop-over site for Wrynecks heading south in the autumn.

ENJOYING WOODPECKERS IN THE FIELD

European species

Once you have seen the British species, where next to watch and learn even more about this fascinating family? Depending on which checklist you follow, there are either 10 or 11 woodpecker species in Europe. You get the eleventh if you accept that Green Woodpeckers are split into two species: Eurasian Green and Iberian Green. Generally, the number of species present in European countries increases as you travel eastwards. For example, there are no woodpeckers in Iceland, only one in Ireland, three regular breeders in Great Britain, seven in Finland, eight in Spain and Sweden, nine in Italy, Switzerland, Germany, Hungary and Estonia, and 10 in France, Austria, the Czech Republic, Poland, Slovakia and the Balkans.

Right: A female Iberian Green Woodpecker. Going on holiday to Spain or Portugal? Then look out for this close relative of the Green Woodpecker that you may know back in Britain.

Further afield

Most of the world's woodpeckers live in South East Asia and South America. The ultimate hotspot is an area that takes in the forests of Colombia, Ecuador and Peru. If you have only seen the European species so far, a woodpecker watching trip to those regions will offer the chance of seeing the smallest, the biggest, and some of the most colourful and spectacular woodpeckers on earth. North America is home to 22 species, some easy to watch as they readily visit backyard feeders. Central America is also wonderful – Mexico, Costa Rica and Panama are picid paradises. Islands often have just a few species or only one, typically unique to that island, like the endemic Jamaican, Cuban Green, Puerto Rican and Guadeloupe Woodpeckers respectively. In the forests of Asia, in the likes of India, Malaysia, Thailand, Indochina, Indonesia, the Philippines and Japan, there are many great woodpeckers, from the miniature to the massive. Africa has some endemics, too.

Above: The Puerto Rican Woodpecker is the only resident woodpecker on the island from which it takes its name. It is also an endemic species so you can't see it anywhere else in the world.

Tips for watching woodpeckers

Visit the right places In Britain at least, head for the woods. Avoid overmanaged plantations and walk where there are old trees and dead wood.

Visit at the right time Woodpeckers can be seen at any time of day, but early morning (when they are out of their roost holes and getting busy) and early evening (when there is often a flurry of activity before they disappear for the night) are often best.

Field skills Tread carefully; avoid rustling leaves and stepping on twigs. Hide your outline by using your surroundings, keeping trees behind you. Do not make eye contact; pretend to ignore the bird.

Clothing You do not have to wear full camo, but dark colours and rustle-free garments will help you get closer to any bird you find.

Listen Learn their calls and drums (there are plenty of resources available). Woodpeckers are often heard before they are seen.

Check all levels Woodpeckers feed at every level, from the canopy to the ground. Always check snags as they love to cling to these.

Be quiet Woodpeckers can be skittish and disappear when humans approach. Chatting, coughing and sneezing alerts them, and all wildlife, to your presence.

Be patient Woodpeckers often hop around the back of trees trunks when they sense someone is watching. Wait a while and they usually hop back into view.

Observe behaviour Once you have found your woodpecker, watch what it does. They are always doing things, and it is rewarding to watch their behaviour.

Go with an expert Joining a guided birding walk with someone who knows where woodpeckers hang out and how they behave can be very satisfying.

Report sightings Make your birding meaningful. Let your local bird recorder, or the BTO, know where you have seen woodpeckers, especially the rare Lesser Spotted Woodpecker and Wryneck.

Left: A Great Spotted Woodpecker peeps out from behind a silver birch in Shropshire, England.

Opposite: Birdwatchers in search of woodpeckers in an old beech forest in Bulgaria.

Glossary

Altricial Chicks that are rather underdeveloped and unable to move around after hatching.

Anisodactyl A foot arrangement with three toes directed forwards and one directed backwards.

Arboreal Relating to, or living in, trees.

Biomimicry The science of copying and using nature's patterns and strategies in technology.

Boreal Relating to the climatic zone south of the Arctic, especially dominated by taiga forests (coniferous evergreen forests).

Coverts Small feathers that cover the bases of larger feathers (upper wing-coverts, upper tail coverts, ear-coverts etc). Coverts help smooth airflow over the body.

Deforestation The removal of forests by human actions such as logging or burning.

Dipterocarp A tall forest tree found mainly in South-east Asia.

Diurnal Active by day. The opposite of nocturnal.

DNA Deoxyribonucleic acid, a molecule that contains the instructions organisms need to develop and reproduce.

Drumming A non-vocal method of communication used by woodpeckers where they beat rhythmically on trees with their bills.

Ectropodactyl A foot arrangement where a climbing bird with four toes rotates its outer rear toe to the side.

Genus (plural **genera**) A group of closely related species.

Irruption An irregular movement of a large numbers of birds, often to areas where they aren't usually found.

Mandible The upper and lower parts of a bird's beak.

Migration Regular seasonal movement of animals from one region to another.

Monogamy The practice of having only one breeding partner.

Monogomous Having only one mate at a time.

Morphology In biology, the study of structure, form and physique.

Moult The process of shedding and regrowing feathers.

Picid The scientific synonym for woodpecker.

Polyandry A system of mating in which a female has more than one male mate.

Polygamy A system of mating in which an animal (male or female) has more than one mate.

Polygyny A system of mating in which a male has more than one female mate.

Ringer A person who attaches individually numbered metal or plastic tags to the leg or the wing of a bird to enable identification.

Scapulars The body feathers that cover the top of the wing.

Taxonomy The science of classification of all organisms and how they are genetically related to each other.

Tridactyl Having three digits on one extremity. In the case of woodpeckers, three toes on one foot.

Zygodactyl A foot arrangement with two toes directed forward and two directed backwards.

Further Reading and Resources

Books

Gorman, Gerard, *The Black Woodpecker: A Monograph on Dryocopus martius* (Lynx Edicions, 2011) A Study on Europe's largest woodpecker, one that has yet to occur in the UK.

Gorman, Gerard, *Woodpeckers of the World: The Complete Guide* (Christopher Helm, 2014) Comprehensive photographic handbook covering all the world's species.

Gorman, Gerard, *Woodpecker* (Reaktion, 2017) Explores the natural, social and cultural history of woodpeckers.

Harrap, Simon, *RSPB Pocket Guide to British Birds: Second Edition* (Bloomsbury Publishing, 2012) A compact, lightweight and informative guide to the most common birds found in Britain.

Holden, Peter and Cleeves, Tim, *RSPB Handbook of British Birds: Fourth Edition* (Bloomsbury Publishing, 2014) A complete, single source of information on Britain's most familiar birds.

Shunk, Stephen, *Woodpeckers of North America* (Houghton Mifflin Harcourt, 2016) Reference guide to the woodpeckers of North America.

Sielmann, Heinz, *My Year with the Woodpeckers* (London, 1959) Classic account of field work and photography with woodpeckers in Germany in the 1950s.

Winkler, Hans, Christie, David A. and Nurney, David, *Woodpeckers: A Guide to the Woodpeckers, Piculets and Wrynecks of the World* (Pica Press, 1995) Detailed handbook to 214 species.

Online

The Woodpecker Network
(www.woodpecker-network.org.uk)
Website and forum set up to encourage and facilitate the study of woodpeckers in Britain and Ireland.

Woodpeckers of the World Facebook group
(www.facebook.com/groups/1438058619755392)
A network of over 5,000 woodpecker enthusiasts who share information and photographs.

Xeno-Canto
(www.xeno-canto.org)
Internet resource of bird sounds from around the world, including most woodpeckers.

Conservation

British Trust for Ornithology (BTO)
(www.bto.org)
An independent charitable research body that combines professional and citizen science to monitor changes in bird populations

International Union for Conservation of Nature
(www.iucn.org)
A non-governmental body that monitors the status of all animal and plant species on Earth.

The RSPB
(www.rspb.org.uk)
The UK's largest conservation charity. Lobbies government, conducts research and manages nature reserves.

The Wildlife Trusts
(www.wildlifetrusts.org)
A conservation charity comprising separate regional trusts and a network of reserves across the UK.

Acknowledgements

I am grateful to everyone in the Bloomsbury Wildlife team who worked on this book, particularly Julie Bailey, Jane Lawes and Alice Ward. Sincere thanks to Peter Powney, who read early drafts of the text and made many helpful comments, Danny Alder who made many suggestions and produced the waveform graphs, and Kyle Turner who kindly supplied his drumming recordings. The following also all helped in various ways: Carl Chapman, Godfrey McRoberts, Graham Clarkson, Jeremy Squire, Peter Lack, Tom Cadwallender, Thomas Hochebner – thank you all. Finally, I'd also like to thank all the birders and ornithologists worldwide who have helped me find, observe and study so many wonderful woodpeckers.

Image Credits

Bloomsbury Publishing would like to thank the following for providing photographs and for permission to reproduce copyright material.

While every effort has been made to trace and acknowledge all copyright holders, we would like to apologise for any errors or omissions and invite readers to inform us so that corrections can be made in any future editions of the book.

Key t = top; l = left; r= right; tl = top left; tcl = top centre left; tc = top centre; tcr = top centre right; tr = top right; cl = centre left; c = centre; cr = centre right; b = bottom; bl = bottom left; bcl = bottom centre left; bc = bottom centre; bcr = bottom centre right; br = bottom right

AL = Alamy; FL= FLPA; G = Getty Images; NPL = Nature Picture Library; RS = RSPB Images; SS = Shutterstock, iStock = iS

Front cover t SS, b Ben Andrew/RS; **spine** Mike Lane/ RS; **back cover** t Frederic Desmette/FL, b Steve Knell/ RS; **1** SS; **3** Kevin Elsby/FL; **4** SS; **5** t SS, bl SS, br SS; **6** SS; **7** SS; **8** Gerard Gorman; **9** Gerard Gorman; **11** tl Imagebroker/FL, m SS, tr SS, bl Matthew Maran/NPL, br Nathalie Houdin/FL; **13** tl SS, tr SS, b SS; **15** tl SS, m Steve Knell/RS, tr SS, b Tony Hamblin/FL; **17** t David Tipling/NPL, bl Roger Tidman/RS, br Richard Brooks/RS; **18** SS; **19** Rosl Roessner/FL; **20** bl Gerard Gorman, bcl Gerard Gorman, bcr komrit tonusin/AL, br Neil Bowman/FL; **21** t Gerard Gorman, b Patrice Correia/FL; **22** t SS, b Gerard Gorman; **23** t SS, b SS; **24** t komkrit tonusin/AL, b Gerard Gorman; **25** Gerard Gorman; **26** l photo researchers/FL, r Steve Gettle/FL; **27** SS; **28** Gerard Gorman; **29** Gerard Gorman; **30** SS; **31** Arterra Picture Library/Al; **32** SS; **33** Dickie Duckett/FL; **34** AFP Contributor/G; **35** t SS, b Emanuele Biggi/FL; **36** t Neil Bowman/FL, b Lizzie Harper; **37** Duncan Usher/FL; **38** t SS, b SS; **39** t Dave Bevan/NPL, b Gerard Gorman; **40** Andy Sands/ NPL; **41** David Williams/FL; **42** Richard Packwood/ RS; **43** t Zdeněk Hýl, b Mathias Ackerknecht, eifel-wildlife.de; **44** t SS, b Stephen Dalton/NPL; **45** Victor Tyakht/AL; **46** Juniors Bildarchiv GmbH/AL; **47** Wendy Conway/AL; **48** t imageBROKER/AL, b Andyworks/iS; **49** t Roel_Meijer/iS, b Yoav Perlman; **50** SS; **51** Gerard Gorman; **52** ImageBroker/FL; **53** t SS, bl SS, br SS; **54** t SS, b SS; **55** t SS, b Mark Hamblin/RS; **56** t SS, b SS; **57** SS; **58** t SS, b iS; **59** John Hawkins/FL; **60** Gerard Gorman; **61** t Roger Tidman/RS, b Richard Brooks/RS; **62** SS; **63** l Rolf Nussbaumer/NPL, r Visuals Unlimited/NPL; **64** ARCO/NPL; **65** Marie Read /NPL; **66** SS; **67** Kevin Sawford/RS; **68** Bill Baston/FL; **69** Evan Bowen-Jones/ AL; **70** t Jack Dykinga/NPL, b SS; **71** Sam Hobson/ NPL; **72** SS; **73** l Andrew Mason/RS, c Evan Bowen-Jones/AL, r David Kjaer/RS; **74** Martin Hale/ FL; **75** Jean-Lou Zimmerman/FL; **76** Dominique Delfino/FL; **77** t Bengt Lundberg/NPL, b Bruno Mathieu/G; **78** Konrad Wothe/NPL; **79** t SS, b SS; **80** Markus Varesvuo/NPL; **81** SS; **82** SS; **83** t Danny Green/RS, b imageBROKER/AL; **84** Colin Varndell; **85** wonderful-Earth.net/AL; **86** Nick Upton/NPL; **88** Danny Green/RS; **89** t Gerard Gorman, c SS, b SS; **90** Frederic Desmette/FL; **91** Kim Taylor/NPL; **92** SS; **93** t Barcroft Media/G, b David Ormerod/AL; **94** t SS, b SS; **95** SS; **96** Gerard Gorman; **97** SS; **98** t SS, b Nick Upton/NPL; **99** t SS, c SS, b John Hawkins/FL; **100** t Gerard Gorman, b Simon Cook; **101** t Robin Chittenden/FL, b Gerard Gorman; **102** Aidan Stock/ AL; **103** Gerard Gorman; **104** SS; **105** Mark Goebel Photo Gallery/G; **106** SS; **107** Michel Poinsignon/ NPL; **108** t De Agostini Picture Library, b SS; **109** Bildagentur-online/Contributor/G; **110** SS; **111** t Pictoral Press Ltd/AL, b SS; **112** Gareth Cattermole/G; **113** Martin Le-May; **114** Richard Brooks/RS; **115** David Tipling; **116** Ben Hall/RS; **117** t Jenny Hibbert/ RS, b Richard Becker/FL; **118** George McCarthy/RS; **119** ViktorCap/iS; **120** Andres M. Dominguez, BIA/FL **121** Gerard Gorman; **122** Mark Sisson/FL; **123** David Hosking/FL; **124** Mike Lane/RS.

Index